Harry Grindell Matthews 1880–1941
Inventor and Pioneer

The Death Ray

The Secret Life of Harry Grindell Matthews

To Ruth and Alex,

Jonathan Foster

Wishing you the very best of everything.

Jonathan Foster
(who once "touched on Sperm"!)

xxx

Inventive Publishing

November '09

Copyright © Jonathan Foster 2008

All rights reserved. No part of this publication may be reproduced or transmitted in any form or by any means, electronic or mechanical including photocopying, recording or any information storage or retrieval system, without prior permission in writing from the publishers.

The right of Jonathan Foster to be identified as the author of this work has been asserted by him in accordance with the Copyright, Designs and Patents Act 1988

First published in the United Kingdom in 2009 by Inventive Publishing.

ISBN 978-0-9561348-0-6

Every possible attempt has been made, by the author, to acknowledge all copyright sources of information and photographs. However there may be instances where the origin of information is not clear. The author does not claim liability for such errors and invites any possible corrections, or omissions, to be brought to his attention via the publisher or the website: www.harrygrindellmatthews.com.

Designed and produced by
The Choir Press, Gloucester

Printed and bound in Great Britain

Contents

Acknowledgements — vii
Introduction — 1

1 New Frontiers — 9
2 The Aerophone — 31
3 Into the Light – *Dawn* — 61
4 Into the Dark – Submarine Detection — 85
5 'You Ain't Seen Nothin' Yet!' — 104
6 'Prometheus' and the 'Death Ray' — 121
7 America and Bankruptcy — 167
8 The Mystery Man of the Mountains — 179
9 Breakfast in London, Lunch in New York — 194
10 The Final Frontier — 214

Appendices
  I.   Time Line — 221
  II. Known Addresses — 229
  III. Patents Filed — 230

Endnote — 232
Further Reading — 233
Index — 236

*With love to Mum and Dad – I owe it all to you,
To Joe who has it all to come,
Those at the Lord Grey School, past, present and future.
Finally to AE and JT – fabulous.*

Acknowledgements

WRITING A BIOGRAPHY is like a treasure hunt and I owe a great debt of gratitude to many kind people who have helped me discover treasures I would have undoubtedly missed had it not been for generosity and support: Ioan Richard for his continued support from the very beginning, Jeff Morgan and his wife Heulwen for their wonderful hospitality and generosity. Eric Garrett for his historical insight, his wife for her kind hospitality and Chris Jones for allowing me to photograph Tor Clawdd. Edna Bodycombe, Mabel Thomas, Ogwyn Norris, Eiryl Davies, Roy and Myra Williams, Helen MacDuff, Frank Jones and Neville Rees for sharing their memories of Matthews. Gari Melville, Andy Gibbard, Graham O'Keefe, Ajit De Silva, for their technical support. All the staff at the Swansea County Archive, National Archives, Kew, British Library, London, The Choir Press, York Publishing Services and the Patent Office. Special thanks to Kate Dutton, and Dr D. Shaw for their willingness and patience to review my work. Thanks to Vanessa Christian who did such a wonderful job on the cover. The author is very grateful to Alan Freke, the owners and residents of 'The Grove' and the Winterbourne Parish Council for all their help with the blue plaque. Carly and Simon for their cosy hospitality and introducing me to some of London's finest curries and finally to a wise old sage called Tim Jones.

Introduction

IN SEPTEMBER 1941 under the malevolent storm clouds of the Second World War a tall, smartly-dressed man looking every one of his 61 years was hurling thick paper files into a roaring bonfire. Those files contained details and patents for wireless radio communication, a boat controlled by a beam of light, submarine detection, the recording of sound on film, a sky projector, an aerial defence scheme to protect cities, liquid fuel propellants, and the infamous 'death ray'. His personal diaries with details of appointments with leading scientists, government ministers and diplomats of the day, together with his notebooks containing meticulously written details of his experiments dating back to the 1890s, were turning to ashes as the smoke drifted into the air and across the Welsh countryside overlooking Swansea. Gazing into the glowing red embers his thoughts glided over a lifetime of inventing: those early days when he returned from the horror of the Boer War full of ideas of how he could make wireless communication a reality, which he did in 1907 – years before Marconi, 'The Father of Radio'. He wistfully recalled that cold, misty morning in December 1915 when government officials witnessed him and his team operate a small boat with a beam of light on Penn Pond in Richmond, London. He still had a copy of the cheque for £25,000, awarded to him for that, on the wall of his study. That money was spent establishing a laboratory in London where he had spent some of the happiest days of his life working on film and sound recording. So far advanced was his camera, which recorded sound and pictures onto the same reel of film, that he couldn't, to his perpetual disappointment, convince the film industry that it would revolutionise the industry. Only later in the 1920s did the film industry realise the possibilities of 'talkies' and adopt the technology. He smiled at the thought of all the wild rumours that spread after his arrival in Swansea, rumours that he was working on a machine to control the weather and stop earthquakes and that he was combating virulent diseases. Having spent his personal fortune on a lifetime of inventing, the penniless inventor, an exile in his own country, was living a frugal existence amidst accumulating debts, unpaid bills and increasing ill-health. He suddenly realised that dusk was falling and with a shiver he turned away from the blazing bonfire and walked back up the bleak grassy mountainside to his small laboratory.

Closing the door against the cold winds he hung up his long overcoat, walked into his study, stoked the cosy fire, and settled down at his desk to attend to his correspondence. He was keen to reply to a letter he had received from an American electrical engineering firm who had offered him a position with them as an electrical engineer. A well-equipped laboratory, regular income, and the warm American sunshine were all very tempting – an offer he couldn't refuse given his current predicament. Reaching for a fresh sheet of writing paper he dipped his fountain pen into the inkwell and wrote the date, September 11th 1941, and began writing his reply when suddenly, feeling a tight pain in his chest, he slumped to the floor, clutching his chest, and lay dying in agony.

Harry Grindell Matthews was one of the most charismatic and intriguing inventors of the twentieth century. Before the end of the nineteenth century he was dreaming of a world where wireless communication was a reality: a dream he realised in a quiet seaside town on September 3rd 1907. At the age of just 27 he filed his first ever patent – *Improved Means for Effecting Telephonic Communication Without Connecting Wires* – his celebrated 'Aerophone'. A pioneer in wireless telephony, he was one of the first people to transmit speech by radio waves, years before Marconi, thereby taking a giant leap in the fledgling technology that would bring continents closer together and create the global village. His Aerophone offered a glimpse into the future for it was the world's first truly portable communication device that would allow people to talk to others whilst in a motor car or travelling on a train. It made Marconi's attempts at wireless communication – he was still only transmitting Morse code – look clumsy and outdated. At the time it was such a sensation that the young Matthews was invited to Buckingham Palace to give a demonstration to Queen Mary. But Matthews, like so many inventors, was truly inept at financial matters and the company he created to market the Aerophone went bankrupt before it went on sale. The costs of developing the Aerophone were enormous. This ground-breaking invention brought him to the attention of the British Government who wanted to see it in action. But during an official demonstration whilst he was setting up his apparatus he discovered officials taking notes and making a close inspection of the apparatus and dramatically called it off, accusing them of trying to steal his secrets. The whole episode led to a deep-rooted distrust that would forever remain between him and the authorities. However it wasn't all bad news, because this proved to be the beginning of the media's love affair with Matthews. Ever willing to champion the lone inventor's cause, he would never be far away from sensational headlines. He achieved a world first with the Aerophone when he spoke to the celebrated pilot B.C. Hucks whilst he was flying. Such was his mastery with radio communication that shortly afterwards he sent the world's first ever press message, from Newport to Cardiff. His next move was to start radio transmissions between England and France. However the British Post Office refused him a licence to build a transmitter. At a time when Marconi was trying to establish radio stations throughout the British Empire, details have now emerged of what has become known as the 'Marconi Scandal'. Members of Asquith's government owned shares in Marconi and saw Matthews as a threat, and so his fate was sealed. It was a body blow to Matthews who was using a superior method of radio transmission which allowed the transmission of speech and not simply Morse signals.

Matthews' work as an electrical engineer was far-ranging: before the slaughter of the First World War was to engulf Western Europe he was exploring the possibilities of both a radio-controlled torpedo and small airborne airship. Collaborating with eminent scientist Fournier D'Albe he invented '*Dawn*', a small waterborne vessel which could be remotely controlled, with a searchlight, from a distance of five miles. The government gave him the astronomical sum of £25,000 for this ground-breaking piece of technology. This episode brought him into contact with two leading figures of the day: Professor Sir J.J. Thomson, the famed discoverer of the electron, and the maverick First Lord of the Admiralty, Lord Fisher, who described Matthews as 'a mechanical man with mechanical vision far more than twice human vision'. Fisher, aware of Britain's precarious defences, wanted the inventor to apply the technology to build an airborne version of *Dawn* to defend Britain against German Zeppelins, but in the end the project was to be shelved.

Through his collaboration with Lord Fisher, Matthews turned his attention to the possibility of submarine detection. After a series of exhaustive trials he was able to detect submarines up to a distance of 12 miles away, long before SONAR technology was available. However one of the experimental trials resulted in a government enquiry into allegations of corruption and spying. The enquiry was a whitewash with the findings never being disclosed, and his submarine detection apparatus was dropped by the government.

Throughout his life Matthews, in his determined, dogged manner never left the columns of the newspapers for very long. He worked with a vision firmly focused on the future despite rejection, setbacks and sheer bad luck. Through his partnership with D'Albe he was even involved in the early days of television when they both worked on a method of transmitting pictures. With a remarkable prescience he foresaw the advent of 'talkies' and in the 1920s invented a movie camera that recorded sound and images onto a standard reel of film, thereby solving the problem of sound synchronisation that had confounded previous inventors. With his remarkable new camera he recorded Sir Ernest Shackleton's farewell speech before he set out on his failed expedition to the Antarctic in 1926. But unable to convince reluctant film magnates that this was where the future lay, Matthews' patents gathered dust until, having lapsed after seven years, the film industry finally saw the potential and adopted the technology – but too late for Matthews to cash in. Later, in 1929, he went to New York to work for Warner Brothers, who had become interested in his research into talking films, and employed Matthews as a consultant to help them develop his ideas of sound recording.

One invention that was to cast a long and dark shadow over all his other achievements was the 'death ray'. During the interwar years, with Britain still reeling from the 'war to end all wars' and nervous about the wave of fascism sweeping Europe, journalists had heard rumours that Matthews had invented an 'invisible ray' which could 'stop a motor working, kill plant life, destroy vermin, explode gunpowder, fire cartridges, and light lamps'. Straight from the pages of science fiction, it was this invention that led to him being called 'Death Ray Matthews' – a title he hated. With the huge press coverage and claims that the 'death ray' could cut out aircraft engines in flight, the Air Ministry took an interest and requested an official test which Matthews, eventually, consented to give. However the government officials announced that they weren't impressed, ridiculed him and professed to be uninterested – but secretly knew he was onto something, offered him £1,000 on condition that he repeat the demonstration on an engine supplied by them and asked him not to enter negotiations with any other third party. But Matthews, totally mistrustful of the government, refused, and said that he was going to France to collaborate with a company who were satisfied with his claims and had interests in his other inventions. Alarmed at all the rumours circulating in the press and with Matthews going abroad with his 'death ray', the government arranged some tests of their own and a small team of scientists set about testing Matthews' claims. Although unable to replicate Matthews' results, the work of that small team of scientists led to the birth of radar, a device that was to save Britain from invasion during the Second World War. Unable to convince a sceptical government of the possibility of his 'death ray', doubts over his claims that he had built a working apparatus continue to this day. However a patent filed by a close colleague of Matthews, at the French patent office, proves the existence of such a device.

During the interwar years the British Government compiled a dossier on Matthews, the files of which are held in the National Archives. Marked 'secret' and subjected to the 30-year disclosure rule, they reveal that he was placed on an official blacklist stating that 'no correspondence or collaboration with him takes place on any matter'. Rejected and reviled by the British Government, Matthews saw opportunities across the Atlantic and emigrated to America. During the five years he spent there he married his second wife and turned his attention to both the 'Luminaphone' and 'Sky Projector'. The Luminaphone was a musical instrument which was played by a series of light beams. The Sky Projector could project images hundreds of feet into the air. Matthews said that, 'The advantages of the machine in war time for conveying coded messages into the air are obvious. It also has many possibilities, for example, for electioneering purposes, for advertising, and also for announcing important items of news.' Nazi officials, including Goebbels, wanted his Sky Projector to beam effigies of the Führer high into the night sky at Nazi rallies, but Matthews refused to hand it over. After sinking a considerable amount of investment into the Sky Projector, commercial success eluded it and with his second marriage heading for divorce, Matthews decided to return to England.

Exiled by the scientific community and vilified by the government, Matthews retreated into seclusion in the Welsh mountains where he built a laboratory surrounded by an enormous electrified fence and continued with his work. There he planned how to defend London against airborne attack from foreign powers. His idea was revolutionary, involving the firing of rockets to a height of 30,000 feet, whereupon aerial torpedoes would release a clutch of bombs that were suspended from parachutes. He also returned to his original work on submarine detection, became a member of the British Inter-Planetary Society and explored the possibilities of using liquid hydrogen to power a rocket-plane, 'The Stratoplane', to a speed of six miles per second.

In 1938, whilst living at Tor Clawdd, (pronounced 'cloud') he married his third wife, Ganna Walska, an opera singer from New York reputedly worth $25,000,000. Making regular visits to the continent to call in on German scientists, and his wife, who lived primarily in France, he was all too aware of Germany's comprehensive re-armament programme and was driven to the point of insanity at the government's apathy and outright narrow-mindedness. They refused to take another look at his work on submarine detection and turned their noses up at his aerial minefield. Shortly after his death, however, government officials raided his laboratory, taking away many of his files and much of his precious equipment.

Since the Second World War Matthews' name has rarely, if at all, been mentioned or publicly acknowledged. The occasional internet article and a long out-of-print biography of him have provided a fascinating glimpse into the life and times of this rather glorious archetypal British inventor. He had always tolerated scepticism and his passing was barely noticed. Those that remember Matthews describe him as quiet and reserved. They recall his intelligence, shyness, modesty and dapper appearance. Photographs of him show him as smartly dressed with neatly combed hair, bow tie, crumpled lab coat and a far-away gaze in his eye, standing in front of masses of electrical equipment. The few articles written about him, mostly on the internet, paint a portrait of the typical eccentric inventor and only add to his image of that of an eccentric crank or charlatan who merely dabbled at inventing. Emphasis is placed on his, ultimately doomed, relationship with the British Government and his failure to successfully demonstrate his 'death ray' to them. But files held in the Swansea Archive and the National Archives reveal the story of a dedicated inventor who never accepted defeat in the face of commercial failure and official indifference. He apparently had no financial backing from any large financial, academic, scientific or military organisation. By the 1990s he was starting to attract some interest, with a Welsh documentary made about him and a small feature on him appearing on the popular BBC series *Local Heroes*. The intervening years had been punctuated by the occasional newspaper article, none of which shed any new light onto his intriguing life. Despite the fact that his life echoes that of so many other famous inventors – John Logie Baird, Sir Frank Whittle and Nikola Tesla, all of whom were years ahead of their time – there is no lasting memorial, no blue plaque and no awards named after him.

Despite the lack of recognition for the contributions he made to wireless communication, his lack of commercial success and an undeserved reputation, Harry Grindell Matthews was still a remarkable inventor, and one of the most enigmatic and shadowy inventors of the twentieth century.

CHAPTER 1

New Frontiers

On March 17th 1880 a young Jane Grindle Matthews gave birth to Harry Grindell Matthews in the sleepy rural village of Winterbourne, South Gloucestershire. It was a long and strenuous birth but nevertheless Harry was a healthy baby. That year also saw Gladstone defeat Disraeli in the general election and take up residence behind the black glossy door in Downing Street for the second time. On the other side of the world the infamous outlaw Ned Kelly was hanged in Melbourne, Australia, whilst over in America the world's first electric street lighting was installed and Thomas Edison was busy testing his new electric railway in his sprawling laboratory at Menlo Park in New Jersey. 1880 also saw an American inventor, Charles Fritts, pioneer a process for the recording of sound on film, a technology with which Matthews would also later have some considerable success, and it was during that year that the word 'radio' was first used in the sense it is used today.

The Matthews family came from a long line of land-owning ancestors with Harry's father and grandfather being yeomen who cultivated acres of farmland and fruit orchards in the picturesque village of Winterbourne. William Matthews, Harry's grandfather, married twice and had two sons, Samuel and Daniel. Daniel, Harry's father, was born in 1849 and went to a boarding school at Weston Grove just outside the historic city of Bath: the 1861 census lists him as a 12-year-old scholar born in Winterbourne. After completing his studies Daniel followed his father into the family farming business and later met and married a local beauty called Jane Rymer Grindle. Shortly after marrying and with Jane expecting their first child, the young couple started looking for a suitable house and eventually Daniel managed to find just the place when in 1870 he signed a lease for 'The Grove', in Winterbourne, a spacious and handsome house, with a good-sized garden, and space enough to accommodate a large family. Shortly after moving in Jane gave birth to a baby girl called Jane, followed by a second daughter, Frances, in 1871. Then around this time, for an unknown reason, Daniel and Jane, along with their two daughters, temporarily stayed with Jane's mother and step-father, at Day House Farm in nearby Tidenham. Jane's parents owned the Ship Hotel, in Alveston, South Gloucestershire, which was close to Mrs Webb's School, where later in 1888 Harry would start school.[1] According to the census for Winterbourne, Daniel Matthews 'derived an income from land'. He was a successful farm manager owning land in and around the Winterbourne area including seven fruit farms and orchards. An astute businessman, a quality that Harry wouldn't inherit, throughout the 1870s and 1880s he expanded the farm, acquiring more land and property. By 1875 Jane was expecting their third daughter, Charlotte. The following year Jane was again expecting another child, but this time it was a son, William. After a difficult birth the young William, a sickly and delicate child, tragically died at the age of just five weeks. Jane was grief-stricken and struggled to cope with the loss of her first son. But in spite of this early tragedy and with the farming business doing well, the couple had another son, Alfred, born in 1878, followed by Harry in 1880. Matthews' first biographer, Ernest Barwell, suggests that he was born at Winterbourne Court, however this remains a matter for conjecture. What is certain is that by 1881 the whole family were living back at 'The Grove' with the census for that year including a young Matthews aged just 10 months. Eliza, the couple's seventh and last child, was born in 1882.

Harry was sometimes called 'Henry' but it wasn't unusual at the time for the names Henry and Harry to be interchangeable.[2] His older sister, Charlotte, would later change her name to Sara. The family was wealthy enough at that time to employ both a domestic servant and a nurse, Martha and Mary Maggs, to help Jane run the large household and look after seven children. It was a life of genteel respectability for the Matthews family, living in a large house, with domestic staff, surrounded by acres of fields, prospering with a large income from a number of farm properties and agricultural concerns.

In 1882, twelve years after they had moved in to The Grove, Samuel, Daniel's brother, took over the tenancy. Why the tenancy was transferred to Samuel is unclear but the entire family moved to Gaunts Earthcott which was another farming village nearby in South Gloucestershire.[3] Sarah Matthews, Samuel's widow, continued to live at The Grove until her death in 1908. Today The Grove is on the busy High Street in Winterbourne and is a residential home for the elderly. In 1883, shortly after giving up The Grove and moving to Gaunts Earthcott, Harry's father died. Little is known about Daniel's death although around the time of his death there was a serious fire that destroyed nearby Winterbourne Court and whether this is connected to his death is not known.[4] How Jane coped with the loss of her husband one can only imagine. She was a loving and devoted mother and Matthews was later to say of her, 'I owe my initiation into the subject I have made my own to that kindly influence and perfect understanding between my mother and myself.'[5] But with a large family to support and a farming business to manage Jane decided after her husband's premature death at the age of 34 to move the entire family to Clifton, a large suburb west of Bristol overlooking the river Avon. The 1901 census shows that Jane was living with her youngest daughter Eliza, then aged 19, and Eliza's husband Thomas Gore, aged 44, at 'The Gratton'. Other members of the family had, by this time, moved away. Harry's elder brother Alfred went on to become a grocer's traveller living at Polygon Cottage in Clifton and would eventually become a director of 'The Park Farms Preserves Ltd' in Winchcombe, Gloucestershire.[6] Harry's elder sister, Charlotte, went on to be a domestic servant.[7] Harry was, in 1901, on active service in the Boer War. After Eliza died, possibly during childbirth, in 1902, Jane and Thomas Gore moved to a smaller house called 'Friezewood' in the parish of Alveston. After retiring from farming Jane Matthews died on September 3rd 1910, aged 61. Thomas continued to live in Friezewood until his death on February 1st 1915, aged 58.

Surrounded by acres of shady, sprawling woods and rolling fields Winterbourne is located on a hill. The River Frome (pronounced 'Froom') meanders through the scenic Frome Valley between Winterbourne and Frampton Cotterell and runs on into the river Avon. Today you can see, rising 100 feet into the air, 11 huge brick arches of the Winterbourne Viaduct striding across the River Frome. Built in 1902, it is the main line for trains running from London to Bristol and on into South Wales. The elegant twelfth-century spire of St Michael's pricks the sky and overlooks the surrounding parish. Under that elegant spire on June 13th 1880, Harry Grindell Matthews was baptized by the reverend A.H. Austen Leigh, a relative of Jane Austen the famous novelist.[8] The sprawling woods and surrounding fields with their sloping meadows and winding rivers would be Matthews' playground where he would spend many solitary hours wandering alone. With no cars, television, wireless sets and the sky free from aeroplanes, the pace of life was far more sedate. New inventions such as the Gramophone were very mysterious contrivances and seen only at fairground parlours where for the price of a penny the listener could hear a rather scratchy, recorded voice. During his solitary rambles in the pleasant Gloucestershire countryside, far away from inner-city deprivations, Matthews developed a life-long fascination and fondness for animals and would, to the consternation of Martha, his nanny, arrive home with a lame cat or distressed bird. One particular incident related by Matthews' first biographer tells how during one of his solitary walks he came across a group of boys teasing a kitten and saw them throw the terrified animal into a pond and gleefully watch it struggle. Matthews, overcome with anger, gathered up the kitten, 'his fist shot out, and the youthful torturer had an unexpected bath'.[9]

In 1888, aged 8, Matthews started his education at Mrs Webb's School in Alveston. Travelling to school using the local horse-drawn bus service, Matthews would later recall those early school days where his 'success was not conspicuous' and he 'was taught the rudiments of the piano'.[10] The school was a private establishment aimed at educating the middle and upper classes and run by Emma Ann Webb,[11] a kindly soul, whose parents ran the first village Post Office. The building still exists but is no longer a school.[12] Matthews was later to say that 'apart from music it must be admitted that I was the dunce of the whole school.'[13] Besides music he would have learned how to read and write. During his early school days he became friendly with Edgar Williams, the local postman, an enthusiastic musician, who taught him the violin and flute – and also gymnastics. Matthews would, in later life, become a talented musician. He also came to know the local village blacksmith, Jacob Savery, whom he thought to be 'the perfect man'[14] and attributed his initial lack of scholarship to him. Jacob lived in Rudgeway and ran a successful blacksmith's shop making 'The Savery Plough' and various agricultural implements. 'A fine figure of a man was the sixty-year-old Jacob; grey, shaggy and strong, and a man with a wise and infinite understanding of boyhood and boyish ideals. It was in those long hours spent in his forge, helping him to the best of my capacity, sharing his bread-and-cheese, that I learned deep truths from his broad and tolerant philosophy; more, indeed, than ever I was taught at school.'[15] It is interesting that the young Matthews had absolutely no scientific education whilst at school. He would later remedy this by enrolling at a technical college in Bristol. In the meantime he would escape the tedium of his schooling by losing himself amongst wires, batteries, electric bells, coils and magnets. A popular book at the time was 'The Boys' Playbook of Science' by John Pepper, a Victorian scientist renowned for giving the most amazing experimental demonstrations, who did more to make science popular and entertaining than anyone else of his generation. The book was stuffed with practical experiments to demonstrate the principles of magnetism, electricity, chemistry, astronomy, mechanics and optics: the list was endless. It was wonderful gift to someone like Matthews who would pore over books to discover all the things a young scientist needed to recreate each exciting experiment. Fuelled with enthusiasm he would pester Martha, collect together and buy the necessary bits and pieces to build a battery, telescope, thermometer, or an optical lantern. Introduced to the wonders of practical science by Pepper, Matthews would later go on to read the classical works of Voltaire, Ampere, Ohm, Galilei and Faraday.

Matthews, being an imaginative child, started wondering how it was possible that birds managed to fly, and his observations of wild birds led him to ponder the phenomenon of flight and how it was possible that they defied gravity. Whilst two American brothers, Orville and Wilbur Wright, were experimenting with methods of achieving flight, the young Matthews was also dabbling in the same arena. After throwing a large piece of board from the open window of a loft and observing the effect of the lift, he concluded that it would be possible to glide with the assistance of some home-made 'wings'. Procuring all the necessary bits of wood, string, fabric and glue, and aided by his older brother Alfred, he secretly built a set of wings in the garden of The Grove. After hours of work the wings were finally completed and ready to fly. Choosing a time when the rest of the family were out and the coast was clear the two small boys climbed into the loft, gingerly carrying the precious wings. Strapping them to his arms, and following in the spirit of Icarus, Matthews stepped out from the open window. Despite his frantic flapping of the flimsy wings, after a few stomach-churning seconds the outcome was inevitable and, although no bones were broken, Matthews later told Jacob of his experiment and how he had come to get the bruises. However this incident, which was to remain clear in his memory for years to come, would not put him off flying as, later in life, Matthews was to become a qualified pilot, owning his own light aircraft. Orville and Wilbur, who had considerably more success than Matthews, went on to achieve the world's first controlled heavier than air flight in December 1903.

Matthews was popular with the other lads in the village but they felt he was 'someone with strange ideas which they could not follow and being unable to understand him, they left him to his own devices'.[16] He was starting to get a reputation as a bit of a loner and an eccentric, interested in science and experimenting. One can imagine him reading, and losing himself in what science literature he could lay his hands on, including popular science fiction of the time, and it wasn't long before Matthews, with his inquiring mind, started some boyhood experiments in the field in which he was later to become a pioneer. By adapting some domestic appliances and spending his pocket money on a few other necessary items he improvised the necessary components to construct a simple transmitter and receiving aerial and with his home-made 'Heath Robinson' apparatus he managed, by means of wireless, to ring a bell over a short distance across a pond.[17] All he needed to achieve this was a simple transmitter, an aerial, an electrical bell and a battery. He would have made a transmitter from two small copper plates placed closely together with only a small gap between them. The copper plates would then be connected to a coil of copper wire, an induction coil, and then to a battery. When the electrical circuit was complete a bright spark, producing a Hertzian (radio) wave, would jump across the gap between the two copper plates. That small spark radiated energy, as an electromagnetic wave, to an aerial connected to the electrical bell, inducing an electrical current, causing it to ring. Hertzian, or radio, waves are named after Heinrich Hertz who proved their existence in 1887. Radio waves are just one type of wave that go to form the many different waves of the electromagnetic spectrum. James Clark Maxwell predicted the existence of electromagnetic waves in 1864. Each type of wave that makes up the electromagnetic spectrum has different properties and therefore different uses, with radio waves being used to transmit sound and pictures. Hertz later discovered that radio waves were reflected off solid objects: the principle behind radar. So even as a schoolboy Matthews was already handling the basic materials that were to shape his own destiny. Astronomy also held a fascination for him and, on cloudless evenings after having read storybooks, he would spend hours gazing at the night sky making observations, mapping the heavens and learning the names of the different constellations. As he peered out from his bedroom window his thoughts would dwell upon the vastness of the universe and what a small part the earth was amongst it all: would it be possible to extend the limit of communication between distant planets?

From an early age Matthews held an ambition to become an electrical engineer, an ambition his mother supported and encouraged. After Mrs Webb's School, he enrolled at the Merchant Venturers' College on Unity Street, Bristol to study electrical engineering. It was here that Matthews would have developed a better understanding of electrical theory and practice. The Merchant Venturers' College was built by The Merchant Venturers in 1865 as a charitable organisation that helped 'to promote learning and the acquisition of skills by supporting education'.[18] The college was an efficient and well-organised institute that would eventually become part of what is now the University of the West of England. The building was closed in the 1960s and after a period of neglect has now been turned into handsome residential apartments.[19]

Little is known of Matthews' time at college but anyone spotting the tall, young Matthews in their class would have no idea that in their midst was a pioneering inventor of the twentieth century. Most of the students were bright and intelligent, though not the intellectual elite, for those had won scholarships to universities, keen to make careers for themselves. Whilst at college Matthews would have studied the history, theory and applications of electricity with a syllabus that may well have included practical classes. However, being away from home for the first time, he did not have a particularly enjoyable time and when he was sixteen, in 1896, he left with a 'greater sense of relief'[20] and started a three-year apprenticeship at a Bristol electrical engineering firm. It isn't known what firm this was but it may have been one of the larger firms, such as The Bristol Tramways and Carriageway Company. With Sir George White, a wealthy philanthropist, as chairman, the company which built and serviced the city's electric tram system was the first of its kind in the United Kingdom.[21] The company also designed and built the rolling stock, tram lines, power cables and the power stations required to generate the electricity to power the trams. Matthews' days as an apprentice were a really happy time: he was 'always the first to volunteer for overtime'[22] and he impressed his employer with his diligence and industry. Although happy and in an environment far more to his liking than college, he would have found an apprenticeship tough going. It was no ivory tower, with long hours spent in noisy, grimy workshops alongside working-class men; no doubt iron would have entered into the soul of the fresh-faced youth from a middle-class background. It was during this period that he learnt far more about the theory of electricity and its practical applications than at college. His diligence paid off, so much so that after a period of 18 months he had learned everything they could teach him and was allowed to cancel his contract and 'accepted an offer from another quarter'.[23] Ernest Barwell, Matthews' first biographer, suggests that at this point he worked with Mr. J.H. Winter who was 'one of the pioneers of electrical lighting'[24] and was able to do far more experimental work than he did during his apprenticeship.

Around the turn of the century, radio and wireless were emerging technologies and people were starting to realise the enormous possibilities. Young Matthews read the scientific journals of the day and followed their development with avid interest. The history of radio and wireless communication and its supporting electronics stretches back for centuries and is crowded with different scientists and inventors all over the world having invented the different components that come together to make the sophisticated wireless devices that exist today. An excellent history of wireless is given in *History of Wireless* by Sarkar *et al.*[25] There is some controversy about who actually invented radio, depending on what literature is read, but it simply cannot be attributed to just one person. Marconi, Lodge, Fessenden, De Forest, Branley, Jennings and Nikola Tesla are but a few of the pioneers who have made important contributions to the evolution of the technology. The American Supreme Court recognised a patent filed by Nikola Tesla in 1897 as the world's first radio patent. However in the same year Marconi obtained a patent, in England, for *'Improvements in Transmitting Electrical Impulses and Signals and in apparatus there-for.'* This was to be the source of much international contention about who invented radio. Marconi would not submit a radio patent in America until November 10th 1900, and it was turned down in favour of Tesla's 1897 patent. However in 1904 the United States Patent Office reversed its decision and gave the radio patent to Marconi. But then two years later the US Supreme Court upheld Tesla's original patent for the invention of radio. The divisions this has caused in the scientific community still exist to this day. Marconi, along with professor Karl Braun, shared the 1909 Nobel Prize for Physics 'in recognition of their contribution to the development of wireless telegraphy'.[26] Braun is more widely remembered as the inventor of the cathode-ray tube which was originally called Braun's Tube.

America was the birthplace of wireless technology and held the lead over Europe for some years. In 1864, an American dentist, Dr Mahlon Loomis, filed the world's first wireless patent describing wireless transmission and thereby ushered in the age of wireless communication.[27] Two years later he managed to transmit signals over a distance of 22km in the Blue Ridge Mountains with his Aerial Telegraph. Amos Emerson Dolbear, an American professor, was granted a patent for a wireless invention in 1882. His device was very rudimentary, to say the least, simply consisting of an induction coil, microphone, telephone receiver and battery. Just the sort of components Matthews would have used around 1890 to ring his electrical bell across that small pond in Gloucestershire. Also in 1882 Nathan Stubblefield, an American farmer from Kentucky, transmitted audio signals without wires. Ten years later in 1892, Stubblefield demonstrated the transmission of speech using a wireless telephone which he patented in 1907.[28] Two Englishmen, Wilson and Evans, demonstrated, on the river Thames in 1888, a radio-controlled boat. Captain Henry Jackson, First Sea Lord of the British Admiralty, was quick to realise the enormous potential of wireless communication for the armed services and successfully sent Morse code signals over a few hundred yards by wireless in 1891.[29] Sir Oliver Lodge demonstrated, at a meeting in Oxford during the autumn of 1894, the transmission of a Morse signal by wireless over 180 feet through two stone walls. Sir Oliver's work had led him to improve the coherer, a device used to detect radio signals, invented by Professor Branly, making it more practical.[30] In 1895, a Russian inventor called Alexander Popoff (or Popov), used the principal apparatus of wireless transmission – an aerial, coherer and an electromagnetic relay – to send and receive wireless signals over a distance of 600 yards.[31] It was in that same year that Marconi transmitted and received a coded message over a distance of one and three quarter miles in Bologna, Italy. But it would be another two years of testing and research before Marconi was able to officially demonstrate a radio transmission of a signal to a small boat over a distance of 18 miles in the Bristol Channel. As the nineteenth century drew to a close, Marconi successfully transmitted an international wireless message from Dover to Wimereux in France.[32] Matthews would have been following these developments with keen interest and knew that this technology would play an increasingly important role in his future career. He never lost sight of his ultimate goal but his life, as with thousands of other men of his generation, was interrupted by the Boer War and in 1899 he found himself leaving Bristol for South Africa.

During his time spent on active service in South Africa, Matthews was to learn much more about the possibilities and applications of wireless communication. With the outbreak of the Boer War (1899–1902), Matthews, aged nineteen, joined the Baden-Powell South African Constabulary (SAC). The SAC, with 8,500 personnel, was formed in 1900 under the leadership of Major General Robert Baden-Powell, the famous scout leader, as part of the British preparations for the policing of captured Boer territory and to establish a permanent British garrison in South Africa.[33] Matthews was posted to Cape Town and then Bloemfontein where, incidentally, the author of *Lord of the Rings*, J.R.R. Tolkien, was born in January 1892. Whilst in Africa on active service he was wounded on two different occasions and decorated for acts of bravery. It was during his service with the SAC that Matthews began to take a more serious interest in the practical applications of wireless communication, more specifically wireless telephony. Wireless telephony is different from wireless telegraphy which is the transmission of Morse signals without wires. Wireless telephony is the transmission of speech and was first demonstrated back in 1892 by Stubblefield. At the time of the Boer War, fortunately for Matthews, South Africa was to play a crucial role in the development of this new, fledgling, technology.

One warm evening in May 1899, the Town Hall in the coastal city of Port Elizabeth, South Africa, was crowded with people to hear a lecture about the work of the wireless pioneer Edward Jennings. In the lecture, which was reported in the *Eastern Province Herald* on May 9th 1899, all the recent developments in wireless technology along with the progress Marconi, Edison and other wireless pioneers had achieved, were discussed. Jennings presented the results of his research and demonstrated his apparatus by transmitting a telegraphic message between two rooms in the Town Hall. Early that same year Jennings had already managed to transmit a signal from a lighthouse over a distance of 13km to Cape Recife. He later gave a demonstration of his apparatus in Cape Town on February 11th 1899. This would have received wide publicity as amongst the many dignitaries present was the Prime Minister of the Cape Colony, W.P. Schreiner. Jennings, born in 1872 was, much like Matthews, never to get the credit for his pioneering work in the field of wireless telegraphy and communications. When Matthews was to later arrive in the region, on active service, he would have heard about these demonstrations and of Jennings' work.

One scientist, above all others, is remembered for his work in wireless technology: Guglielmo Marconi. Born in Italy in 1874 to wealthy parents, Marconi is considered by many to be 'The Father of Wireless'. Whilst studying at college, in 1894, he became interested in the idea of using radio waves as a means of communication at a time when messages were transmitted over telegraph wires. In the true sense of an amateur scientist he set up his own laboratory in the attic room of his parents' house and began experimenting with the most rudimentary equipment. Within less than a year he had managed to send and receive radio waves over a distance of almost two miles. Flushed with this remarkable achievement the young inventor took out his first patent in 1896. Unable to interest the Italian Government in his work, he approached the British Government and suggested their armed forces might find a use for his invention. Britain, with more foresight than Italy, thought that wireless telegraphy could be used by the armed forces in South Africa and invited Marconi to demonstrate his wireless equipment. Engineers working for Marconi made a successful demonstration of wireless Morse in Cape Town on December 4th 1899. Earlier that same year Marconi set up a small wireless station at Wimereux in France and after refining his equipment he was able to transmit a signal across the English Channel – a distance of 31 miles.

However further tests in the field proved less successful due to adverse weather conditions causing problems with the aerials and the kites used to hold the aerials and transmitting wires aloft, problems that would later dog Matthews and his team. In the end Marconi's wireless sets weren't used but, having witnessed successful trials of the equipment the year before, the

Royal Navy asked for the wireless sets to be installed on five of their cruisers. Marconi's equipment proved to be very practical, allowing Royal Navy ships to remain in contact over much larger distances than before, bringing obvious benefits to the arena of warfare. The Institute of Electrical and Electronics Engineers (IEEE) reported Marconi's achievement:

> FIRST USE OF WIRELESS TELEGRAPHY
>
> The first use of wireless telegraphy in the field occurred during the Anglo-Boer war (1899–1902). The British Army experimented with Marconi's system and the British Navy successfully used it for communication among naval vessels in Delagoa Bay, prompting further developments of Marconi's wireless system for practical uses.[34]

On a blustery night on December 12th 1901, engineers working for Marconi transmitted the Morse code signal for the letter 'S' from Poldhu, Cornwall across the Atlantic Ocean to Newfoundland, Canada, where, listening intently into his headphones, a jubilant Marconi heard the unmistakable dot, dot, dot. To this day there is some doubt and scepticism as to whether Marconi actually heard Morse signals or just static interference. Indeed the only evidence of this event is Marconi's own entry in his laboratory notebook held in the Marconi Company archives.[35] But nevertheless it was a courageous experiment and international wireless communication had, in the eyes of the public, truly arrived. It is interesting to note that a system of wireless communication using electrostatic induction was invented by Edison and later sold to Marconi in 1903.[36] Edison had already dabbled in wireless communication when as far back as 1885 the extrovert American inventor patented a system for wireless telegraphy between ships and shore stations.[37] Not until much later was the practical value of ship-to-shore communication demonstrated. In 1910 the captain of the SS *Montrose* sent a wireless Morse code message to England informing the authorities that Dr Crippen, then wanted on suspicion of murder, was on board. When alighting at Montreal the unfortunate doctor found a warm welcome from the authorities.

Like Matthews, Marconi's genius lay in his practical rather than a theoretical approach to technical problems, and his ability to organise and bring together the achievements of others. Events on the night of April 14th 1912 were to secure Marconi a place in history. For it was on that night that RMS *Titanic*, after colliding with an enormous iceberg, was sinking in the freezing waters of the Atlantic Ocean. It would be the Morse code distress signals sent out using Marconi's wireless telegraphy system that brought ships in the area to the rescue, saving hundreds of lives. Without a doubt Marconi was a true wireless pioneer and laid the foundation stones of international wireless communication. A comprehensive account of his life and achievements can be read in Gavin Weightman's *Signor Marconi's Magic Box: How an Amateur Inventor Defied Scientists and Began the Radio Revolution*.

So wireless communication had made great strides during the Boer War and Matthews would have had a front row seat. Without a doubt wireless communication was revolutionising the way armed conflicts were being fought. Watching these developments with interest Matthews asked himself, 'If it's possible to transmit Morse signals through the air without wires then surely the same can be done for the human voice?'

In the spring of 1902, just before the Treaty of Vereeniging was signed ending the Second Boer War, Matthews was invalided home as a result of enteric, or typhoid, fever, which impaired his health for the rest of his life, and was awarded a life pension.[38] Arriving back in England, tired and in fragile health but relieved to have survived, he was keen to press on with his ideas for wireless communication. He took some cheap lodgings in Bexhill-on-Sea, in East Sussex, and began to look for what opportunities there were for an electrical engineer and eventually, after calling on the wealthy aristocrat the 8th Earl De La Warr, he managed to secure a position as a consulting engineer for the Earl. Matthews had previously met the Earl whilst he was on active service in South Africa and the Earl was employed as a war correspondent for *The Globe* magazine. Gilbert Sackville, the 8th Earl De La Warr (1869–1915), was a wealthy patron and, amongst many philanthropic gestures towards Bexhill-on-Sea, in 1896 built the Kursaal, an elaborately ornate pavilion 'for refined entertainment and relaxation' to promote Bexhill-on-Sea as a new up and coming seaside resort.[39] Two years after it opened it was the first venue in Bexhill-on-Sea to show moving pictures, and from 1900 onwards this novel form of entertainment became a fixture of the Kursaal's entertainment programme.[40] Moving pictures was an interest that Matthews would later return to when he solved the problem of simultanously recording sound and images. Shortly after marrying Muriel Brassey in 1891, the Earl spent what was for the time the enormous sum of £15,000 on his home, the 'Manor House', which was fitted with electricity and a telephone.[41] Unfortunately, after one of his business partners went bankrupt he had to pay out thousands of pounds to various business concerns and, as a result, suffered considerable damage to his reputation and personal finances. The Earl had a roving eye and a roguish temperament and around this time he committed an act of infidelity against his wife who, in light of his financial troubles, decided to forgive him.

After being wounded during an attempt to rescue an injured soldier he returned home from the Boer War in June 1900 to a hero's welcome. However he soon fell back into his roguish ways and upon his return to Bexhill-on-Sea began a passionate affair with an actress who, at that time, was working at the Kursaal.[42] With *The Times* reporting details of his divorce in July 1902, including a letter written by the Earl confessing to his act of infidelity, his private life became the hot topic of local gossip. As a result of these scandalous revelations the Earl's ambitions to become Mayor of Bexhill were thwarted. But in spite of these setbacks he later went on to marry again and became mayor in November 1903.[43] The whiff of scandal, never far away from the unfortunate Earl, descended upon him again when in 1914 his creditors were again chasing him for money and his second wife divorced him for adultery and desertion.[44] As a result of the ensuing scandal, he bought a commission and he left the country to serve as a major in the Royal Sussex Regiment. He was later given command of a motor cruiser whilst serving in the Royal Naval Volunteer Reserves. When he was stationed in Messina, on the island of Sicily in Italy, he contracted a severe fever and died on December 16th 1915.[45]

With a meeting of minds and the deep pockets of the Earl, Matthews set up a small laboratory and radio station in the Kursaal, and began trying out the ideas he had had during those long evenings in South Africa. What Matthews wanted to do was repeat the experiments of Marconi and Jennings and to extend the range of wireless communication. But Matthews thought the transmission of speech, not Morse code, was where the future lay, and his ultimate goal was the transmission of speech by radio waves. He made numerous trips to London to meet with several leading scientists in the field and seek their advice but he was told that his ambition of transmitting speech without the aid of wires was an 'impossibility' and that 'his current research would take him down a dead end'.[46] This is rather surprising given the recent advances made in both wireless telegraphy and telephony during the early 1900s. Marconi thought, at the time, that there was no real commercial value in long-distance radio broadcasting and that radio transmission of speech over great distances wasn't possible. But the optimistic Earl De La Warr encouraged the young inventor to continue with his work and about five years after starting work for the Earl, Matthews realised his dream of wireless communication when on September 3rd 1907 he transmitted speech by radio waves.[47] That first historic broadcast was made from the roof of the Kursaal 'where he succeeded in establishing a radio station'.[48] Matthews used a small motor boat to conduct some of his experiments off the East Sussex coast and after much trial and error 'succeeded in establishing wireless telephonic communication between points of half a mile'.[49] The use of a boat enabled the distance between the transmitting apparatus, in the Kursaal, and the receiving apparatus, on board the boat, to be easily altered, quickly allowing Matthews to determine the range of his wireless broadcast. Although only a short distance it was a quantum leap forward and showed the potential of voice transmission. After all, in its early days Morse code was only transmitted over very short distances, but by this time it was being transmitted across the Atlantic. Matthews wasn't working on the same lines as those before him, he was working on parallel lines. He enjoyed his time working for the enthusiastic Earl De La Warr, where he would remain until 1909, because it gave him the freedom and time to develop some of his ideas on wireless telephony.

Through his regular trips to London Matthews kept a close eye on what other ambitious inventors, including Marconi and

Fessenden, were doing. Reginald Aubrey Fessenden, a Canadian inventor and a sceptic of Marconi, was another interesting pioneer of radio. After realising the limitations of the coherer, a fragile and unreliable device, he invented the barretter, a device that was infinitely more sensitive to radio waves. Fessenden achieved the distinction of being the world's first radio broadcaster when he transmitted speech over a distance of 25 miles in 1900.[50] His work in wireless telephony led him to develop the synchronous rotary-spark-gap transmitter that generated a continuous wave signal which was more effective in carrying speech. On Christmas Eve 1906 he did the world's first radio broadcast of speech and music.[51] Later in life he went on to invent SONAR. The works of Fleming and De Forest, too, should not be overlooked. Sir John Fleming, an electrical engineer who studied under the celebrated James Clerk Maxwell, invented the oscillating valve in 1904. Also referred to as the vacuum-tube diode, this small device converted the minute electrical oscillations within a receiving aerial into a direct current that could operate the radio receiving apparatus and be heard through a set of headphones.[52] Fleming was a close associate of Marconi, for whom he worked as Scientific Advisor, a role he previously had when he worked for the Edison Company. In 1906 Lee de Forest replaced the oscillating valve by inventing the vacuum-tube triode, also known as the audion, which gave a much louder signal by bringing about better amplification of the incoming radio signal. The triode made long-range telephony possible and was not replaced until many years later with the arrival of the transistor.[53] Matthews would have been aware of the ground-breaking milestones being reached by his contemporaries and his work in a genteel seaside town on the south coat of England proved that he was on the right lines.

Things were also happening in Matthews' personal life, for in 1904 he married Katy Beatrice Williams. Although working in Bexhill-on-Sea, Matthews regularly commuted to visit his mother who was living in Rudgeway, South Gloucestershire. He had been staying with some friends, the Williams family, at Churchill House in nearby Olveston when he met the young and attractive Katy. Katy Williams lived with her widowed mother and worked as a barmaid at the White Hart public house with her grandfather, Edwin Organ, who was the Landlord.[54] After marrying Katy, Matthews continued to work for Lord De La Warr but would make regular visits home to be with his young wife. There are no records showing that they ever lived together – indeed little is known about his first marriage, with no reference at all made to it by Matthews' first biographer – but it was not destined to be an enduring one for it was dissolved a few years later. It would have been difficult for Katy, who wanted to remain in Olveston with her family, with Matthews being away in Bexhill-on-Sea for long periods. In addition, like many inventors he was quite obsessive and preoccupied with his work, all of which would have made for a difficult marriage.

One of Matthews' strengths was that he could see that wireless technology wouldn't just revolutionise communications but also had many other applications. Reports from around this time refer to experiments where he was able to control a small boat using radio waves over a distance of three miles.[55] But ever the pioneer he developed this idea with a practical end in mind and constructed a radio-controlled torpedo device that could operate under water. An American engineer, Harry Shoemaker, had designed radio-controlled torpedoes in the United States back in 1905.[56] The *Evening News* of April 9th 1908 reported that Matthews had been working for a number of years on a device that 'can be steered to the left or right. Its engines can be stopped completely; the torpedo brought to a standstill for any length of time, and then restarted at the will of the operator.' The device was

> a simple apparatus ... its essentials are at the transmitting station, a source of electricity, such as an accumulator, a spark coil, control keys, a selective device, and a wave transformer and transmitter. The outfit weighs about 200 lbs. In the shell of the torpedo is placed the sensitive mechanism which receives the wireless power from the sending station and uses it for the purpose of the operator, who may be a mile or two distant.[57]

The actual operating distance was nearer eight miles but, not for the first time, Matthews played down his achievements when dealing with the press. What was so revolutionary about this device was that it didn't require an aerial, and the standard wireless equipment on board a battleship would not interfere with its operation.

> The operator sits at a row of small keys, and by the position of his selective device he can tell exactly how to steer the torpedo. By day the wake of the torpedo would tell him the course, and his control keys would give absolute command of it. At night a tiny electric lamp, carefully screened from the enemy, would be used to show the position of the torpedo, and by this it would be guided with absolute certainty until the fatal blow was struck.[58]

Conventional torpedoes used by Germany in the First World War had a range of about 800 yards.[59] However Matthews' wireless torpedo had a controllable range of just less than eight miles.[60] Torpedo nets were used at the time to protect large naval vessels. These heavy steel nets hung over the side of a moored ship and provided effective defence against conventional torpedoes. When a torpedo struck the net it would explode at a safe distance from the hull thereby avoiding any damage. Matthews proposed the use of a conventional torpedo to strike a hole in a defending torpedo net and then using the wireless-controlled torpedo to steer through the gap left in the net by its predecessor. A torpedo that could be steered whilst in motion could also be used to disable the rudder or propellers of an enemy battleship. The *Black and White* magazine reported that, 'The inventor has demonstrated the capabilities of the device under the most severe conditions, and has abundantly succeeded where others have failed.'[61] Matthews suggested that a mast be fitted to enable the progress of the torpedo to be followed from the conning tower of a submarine.

During his wartime service, Matthews would have handled and gained knowledge of explosives and the *Mall Gazette* of September 3rd 1907 gave details of how he had constructed a device that could 'act as a detonator exploder by connecting a detonator to the receiver'. When the receiver picked up the radio waves it would activate the detonator and detonate the explosive. Designed to be portable, the explosion could be remotely controlled via radio waves. Such a device could be used to detonate bridges and railway lines without the need for trailing wires between the explosive and detonator.

Articles about Matthews' work started appearing in the press and thus began a long, curious and rather ambiguous relationship with Fleet Street. The *Bexhill Chronicle* of April 6th 1907 reported details of his work on the wireless transmission of speech. The *Chronicle* reported details of his 'Telephone Without Wires': 'Mr. Matthews has succeeded in establishing communication through the Bristol Channel, a distance of $7^{3}/_{4}$ miles.' Interestingly the reporter goes on to say, 'In order to escape the unwelcome attention of certain foreigners interested in the matter, Mr. Matthews will shortly be leaving the town.' The press was, in the main, on his side throughout his long career and he would later use his good relationship with them to his advantage. But it would be the misleading reports about his 'death ray' that would tarnish his credibility in the eyes of both the public and the British Government, in a few years to come.

By the beginning of the twentieth century wireless technology had well and truly arrived. The technology was to play a crucial role opening new frontiers in both the twentieth century and in the life of a young inventor busily working in a genteel seaside town on the south coast of England.

Notes

1. Information kindly supplied by Rosemary King of the Alveston Historical Society.
2. Conversation with Mr Eric Garrett, 31/03/07.
3. Information kindly supplied by Rosemary King of the Alveston Historical Society.
4. Conversation with Mr Eric Garrett, 31/03/07.
5. Barwell, Ernest H.G., *The Death Ray Man: The Biography of Harry Grindell Matthews, Inventor and Pioneer* (Hutchinson, 1943), p. 14.
6. Swansea County Archive, D/DZ 346/10.
7. www.frenchaymuseumarchives.co.uk/Archives/Elliott/07_Older_Houses.rtf (accessed 04/01/07).
8. Ibid.
9. Barwell, p. 15.
10. Ibid.
11. Conversation with Mr Eric Garrett, 31/03/07.

12. Information kindly supplied by Rosemary King of the Alveston Historical Society.
13. Barwell, p. 15.
14. Ibid.
15. Ibid.
16. Ibid., p. 16.
17. Ibid., p. 17.
18. www.merchantventurers.com/society.htm (accessed 17/09/06).
19. www.10unitystreet.com/history.html (accessed 17/09/06); www.merchantventurers.com/default.htm (accessed 17/09/06); www.uwe.ac.uk/aboutUWE/history.shtml (accessed 17/09/06).
20. Barwell, p. 18.
21. Conversation with Mr Eric Garrett, 31/03/07.
22. Barwell, p. 18.
23. Ibid.
24. Ibid.
25. Sarkar, T.K. *et al.*, *History of Wireless* (John Wiley & Sons, Inc., 2006), p. 282.
26. Ibid.
27. Ibid., p. 68.
28. Ibid., p. 77.
29. Ibid., p. 84.
30. Ibid., p. 86.
31. Ibid.
32. Ibid., p. 92.
33. www.roll-of-honour.com/Regiments/BoerWarNotes.html (accessed 13/03/07); www.warmuseum.ca/cwm/boer/southafricaconstabulary_e.html (accessed 13/03/07).
34. Sarkar, p. 451.
35. Belrose, John S., 'Fessenden and Marconi: Their Differing Technologies and Transatlantic Experiments During the First Decade of this Century', available at: www.ewh.ieee.org/reg/7/millennium/radio/radio_differences.html (accessed 02/01/08).
36. Sarkar, p. 79.
37. Ibid., p. 255.
38. *York Gazette*, August 3rd 1912.
39. www.bexhill-on-sea.org/historyofbexhill.php (accessed 12/07/07).
40. Porter, J., *Bexhill-on-Sea. A History* (Phillimore & Co. Ltd, 2004), p. 86.
41. Ibid., p. 66.
42. Ibid., p.73.
43. Ibid., p. 80.
44. Ibid., p. 94.
45. Ibid.

46. Barwell, p. 19.
47. Swansea County Archive D/DZ 346/9.
48. Barwell, p. 19.
49. Ibid.
50. Sarkar, p. 92.
51. Ibid., p. 409.
52. Ibid., p. 98.
53. Ibid., p. 101.
54. Conversation with Mr Eric Garrett, 31/03/07.
55. Barwell, p. 19.
56. Sarkar, p. 100.
57. *Black and White*, April 3rd 1909, Swansea County Archive D/DZ 346/5.
58. Ibid.
59. Willmott, H.P., *World War I* (Dorling Kindersley, 2003).
60. *Black and White*, April 3rd 1909, Swansea County Archive, D/DZ 346/5.
61. Ibid.

CHAPTER 2

The Aerophone

O<small>N NOVEMBER</small> 6<small>TH</small> 1909 Matthews filed his first ever patent, *'Improved Means for Effecting Telephonic Communication Without Connecting Wires.'*[1] His address is given as Friezewood, Rudgeway, where he was lodging with his mother at the time, and his occupation as Electrical Engineer. The patent gives details of *'improved means for effecting speech communication by means of transmitting and receiving instruments without the aid of connecting wires the installation being of portable character.'*[2] This invention was Matthews' much celebrated Aerophone device, and announced his entry into the world of wireless technology. Having worked under the patronage of Earl De La Warr since his return from the Boer War Matthews had designed and constructed a portable radio telephone. But this wasn't the only patent Matthews filed that year, for one month later, in December, he filed a patent for an *'Improved Automatic Righting Device for a Flying machine.'*[3] This device enabled a flying machine to 'automatically right itself to its normal vertical position when subjected to an undesired temporary displacement relatively to a vertical line'. This was a truly remarkable device that consisted of an electric circuit, revolving motor, electromagnets, and a gravity control switch. The device was connected to the aerofoils of a biplane by a system of levers, pulleys, a fly wheel and connecting wires. Via the action of a mercury switch the aerofoils would automatically compensate should the aircraft deviate from the normal level. Matthews installed his automatic pilot device on a Farman biplane and invited the well-known aviator of the time, Mr Grahame White, to test his automatic pilot.[4] With diverse interests covering electronics, wireless technology and aeronautics, along with two patents under his belt, Matthews was gaining a reputation as a gifted and versatile inventor. File Air 5/179 held in the National Archives shows that a Mr W. Stocken of Kensington had financed Matthews from 1908 for a period of five years. Exactly who Mr Stocken was remains a mystery but he would, in all probability, have been given a stake in Matthews' Aerophone in return for his patronage. Matthews' finances were always rather precarious and he would often, and quite unjustly, be referred to as 'Grindell the Swindle'. He never ever set out to deceive or swindle anybody, always offering a fair stake in his patents. However the inventor and the entrepreneur are worlds apart, with the inventor happy in his workshop oblivious to money and the entrepreneur in his counting house. In the early days Matthews was financed by his family's estate, Earl De La Warr and Mr Stocken.

News of what Matthews had been doing with his earlier work on wireless communication had reached Sir Joseph Lyons. Sir Joseph was a wealthy entrepreneur who had amassed a fortune building, along with his partner Montague Gluckstein, the Lyons Empire which included a chain of popular tea shops, up-market restaurants and several hotels. Sir Joseph was himself an inventor and one of his more successful inventions was the chromatic stereoscope which was a device that combined a telescope, microscope, magnifying glass and binoculars. In the 1930s the Lyons catering empire introduced the world's first automated machine to process the millions of transactions carried out by the company and the Lyons Electronic Office (LEO) was born. Built by the company's own engineers, LEO was one of the world's first business computers and was used to calculate such mundane things as rates of pay, hours worked and stock levels.[5] When the two met Sir Joseph enquired how far Matthews could transmit voices using his apparatus and after some discussion they arranged a test whereby Matthews would transmit his voice from 'a private room in the Shaftesbury Avenue building, where the Trocadero restaurant was to be found, to the smoke-room of the Eccentric Club on the opposite side of the avenue.'[6] No longer under the patronage of the 8th Earl De La Warr and with limited financial resources, according to Barwell 'the young inventor had that morning pawned his overcoat to enable him to transport his instruments to the place of the demonstration'.[7] However Sir Joseph was impressed with what was apparently a very successful test: in fact so impressed that he was prepared to put up £500,000 'for the development and exploitation of the invention' if, and this was a very big if, he 'could hold a radio-telephone conversation between my firm's headquarters Cadby Hall in Kensington, and the Trocadero restaurant' in London's Piccadilly Circus.[8] This is a distance of nearly four miles across London. But Matthews knew that this was beyond the capabilities of his Aerophone at that time and was bitterly disappointed at not being able to collect such a fabulous prize. Although he had previously established wireless communication over a distance of 7¾ miles across the Bristol Channel, doing the same across a crowded city, with all the obstacles, was a different matter. The technology was simply too limited and this feat of science lay some way in the (not-too-distant) future.

Returning to his London hotel later that evening and contemplating how he could continue financing his experiments to extend the reach of the Aerophone, Matthews struck up a conversation with a small party of men gathered in the lounge. He began to relate his story of how he had invented a device which enabled people to speak to each other without the aid of wires and had just demonstrated it to a wealthy entrepreneur who had issued him with a challenge and a prize of half a million pounds should he be successful. But his current financial situation was so dire he couldn't afford to accept the challenge. Discussing the obvious commercial possibilities of wireless communication and what the future might hold for such technology, his audience became captivated. Matthews enthusiastically explained to that small gathering just how his remarkable invention worked. When the operator wanted to transmit a message he simply pressed a button which activated the battery and telephone transmitter. Not requiring a conventional aerial, the message was transmitted and received at very high frequency through a series of coils of insulated wire wrapped around bundles of iron wire. In preparation for receiving the transmitted message on the second unit, the operator switched off the battery to the transmitter circuit and switched on the telephone receiver and relay circuits. The receiving unit would emit an audible sound prior to the incoming signal. By alternating between the transmitting and receiving circuits a conversation could be held over a distance. The audience listening to the young, impoverished inventor included Sir Clifton Robinson, Sir William Bull MP, Sir Frederick Frankland, Bart, Mr F.R. Poole and Mr John Cameron who were all suitably impressed and went on to become directors of 'The Grindell Matthews Wireless Telephone Company'. Formed on April 27th 1910, with Matthews as 'Superintending Expert' on a salary of £500 per year, the company, with its head office in Broad Street, London, was set up to market and develop the Aerophone which Matthews had patented back in 1909. To qualify as a company director you had to hold no less than 100 shares and each director would receive a remuneration of £75 each year. The board decided that the company would acquire the patent for the Aerophone from Matthews for £6,666, payable as £1,000 in cash and £5,666 in shares.[9] It was also decided amongst the directors that the company would have a stake in any other invention of Matthews' relating to wireless telephony. Once perfected, the company proposed to sell the patent for the Aerophone to a larger company. Sir Clifton Robinson, a JP, was chairman of the newly formed company, which was set up with a nominal capital of £25,000, in £1 shares. The total amount of capital invested into the company by its shareholders was £10,562.[10] Sir Clifton had an enviable reputation as an engineer of trams and was the chairman of Imperial Tramways Company. Matthews may have had a stroke of luck here for Sir Clifton had connections with the engineering company dealing with the tramway network in Bristol, where Matthews may well have served his apprenticeship. A workshop for Matthews was set up in Pilning, South Gloucestershire, on the banks of the River Severn, and a motor boat, *Undine* was also purchased to provide a floating workshop. Meaning a 'female water-spirit', *Undine* was a narrow boat: 32 feet in length, 7 feet in breadth and 3 feet deep. Powered by a small motor giving the vessel 13 BHP, it was installed with all the necessary wireless apparatus and used as a mobile radio station. *Undine* was later sold to a Mr Arthur Burley, a boilermaker from Chepstow, for the princely sum of £70 in January 1913.[11] So

Matthews' immediate financial problems were solved with the formation of a company which gave him all the necessary facilities, resources and freedom to continue his research and realise his ambitious plans. What Matthews wanted to do was not only to perfect and manufacture his Aerophone for the mass market – he wanted to establish a radio station capable of wireless transmission over vast distances.

With the massive cash injection provided by The Grindell Matthews Wireless Telephone Company the development of the Aerophone was set apace and Matthews was able to give a demonstration to the directors of Messrs Wernher, Beit and Co. on April 13th 1910 in their London Offices.[12] A wealthy firm that dealt in diamonds Messrs Wernher, Beit and Co. was owned by Sir Otto Beit and Julius Charles Wernher. Sir Otto Beit was a Fellow of the Royal Society, a wealthy financier and philanthropist, Julius Charles Wernher owned a diamond mine, was also a financier, philanthropist and collected fine art. Matthews was 'asked to hold a conversation from inside the strong-room which contained millions of pounds' worth of diamonds. The strong-room was constructed of steel-armoured plating, firebricks and concrete.'[13] From this strong-room Matthews had to broadcast to the other offices in the same building. The demonstration was a roaring success and so sensitive was the apparatus that when he placed his watch close to the microphone, listeners in offices at some distance away were able to hear it ticking. The gathered party warmly congratulated the inventor and eagerly discussed the obvious commercial possibilities of the Aerophone and what it would mean to businesses, offices and banking houses around the world. The complete apparatus consisted of two handmade, highly polished mahogany boxes, nine inches square with brass fittings, housing the various electrical components along with both the transmitter and receiver, in fact everything one needed to hold a conversation at a distance without the need of trailing wires and cumbersome aerials.

On May 2nd 1910 an excited audience gathered at the London Hippodrome watched in awe a wireless-controlled flying model airship. Operated by Mr Phillips, an electrical engineer from Liverpool, who theatrically claimed 'I can sit in an armchair in London, and make my airship drop a bunch of flowers into a friend's garden in Manchester, Paris, or Berlin.'[14] The Hippodrome was packed night after night with the audiences being dazzled by the spectacle of this engineer controlling a small airship at will, and all without wires. Vivid accounts were printed in the press where 'to the accompaniment of tremendous applause, he sent the airship sailing slowly over the auditorium, released a flight of paper birds from its hold, made it manoeuvre in mid-air, and finally brought it back to the stage'.[15] Mr Phillips claimed to have spent approximately £5,000 over a two-year period in designing and building his airship. Matthews read the colourful newspaper reports and decided to go and see the show for himself. He was very impressed by what he saw and managed to get an introduction to Mr Phillips. The *Daily Express* reported that 'Mr. Grindell-Matthews warmly congratulated the inventor on his success, and they plunged at once into a conversation bristling with technicalities.'[16] After speaking to Mr Phillips, Matthews thought that it would be possible to use some of his own wireless apparatus to take control of Mr Phillips' airship and issued a challenge by writing a letter to the editor of the *Daily Express*:

> I most heartily congratulate Mr. Raymond Philips on his success and should like the opportunity of meeting him to determine whether my instruments will deflect the course of his aerial craft. ... I myself have for years been experimenting on a similar line ... with all possible deference and respect to Mr. Raymond Phillips ... I can point out a flaw in his invention. I believe that it would be possible for another operator to interfere with Mr. Raymond Phillips' control. That is to say, assuming that he were to send his airship over my head with a view to dropping flowers upon me, I think I could by manipulating an instrument of my own, compel it to turn around and return to a place from which it was sent. Whether I am able to do this or not I should be only to glad of an opportunity to try and I should like nothing so much as a sporting contest – a wireless duel, so to speak – with an inventor whose genius has filled me with admiration.[17]

Upon reading the challenge issued by Matthews Mr Phillips replied 'I do not say that another person could not intercept a wave sent to my airship but, unless he knew my code I fail to see what he could accomplish.'[18] By 'code' he was referring to the wavelength of the radio wave which he was using to operate his model airship. Mr Phillips wrote to the *Daily News* the next day to accept the challenge. However the story was to take an intriguing twist at this point. The management of the Hippodrome, according to the *Daily News*, May 4th 1910, stated that 'electric currents are dangerous' and that they had 'no desire at present that such an experiment should take place within its walls'. Mr Phillips was, by all accounts 'looking forward to the contest with breezy eagerness'. But

> he did not deny that within the confined space of a music hall it might be possible to tamper with his electric waves. 'With the airship which I am now demonstrating it is quite possible that Mr. Matthews may be able to intercept the waves in such a small space. But as a secret code is used for the transmission of all signals it is likely that my acceptance of the challenge may have the most interesting results.' (Author's italics.)

Clouds of doubt were descending on Mr Phillips and his model airship.

So despite the reservations expressed by the manager, Mr Oswald Stoll, of the Hippodrome, he invited Matthews to build and demonstrate a miniature airship of his own and pitch it against Mr Phillips'.[19] The winner of the 'duel' would receive £250. So the challenge now wasn't for Matthews to interfere with Mr Phillips' control of his airship but to build one of his own and pitch it against Mr Phillips'. Mr Stoll wrote to the editor of the *Daily Express* on May 6th 1910:

> *Everybody knows that Mr. Raymond Phillips airship can be deflected directed and controlled we do not want Mr. Grindell Matthews to show us that.* [Author's italics.] What I want to see is another airship, produced by Mr. Grindell Matthews. Then we can arrange a sham fight upon the conditions laid down by somebody capable of setting them out, and the victor of the sham fight will be the winner of the £250 which I offer.

Members of the Aero Club and an expert from the British Military services would be present to adjudicate and see fair play. Matthews replied saying that

Mr. Oswald Stoll has twice proved my original contention by admitting that the Raymond Phillips airship is not a practical weapon of aerial warfare – that is to say, he admits that anyone can interfere with the operator's control of it.' ... 'This does not agree with the original statement of Mr. Raymond Phillips, who told me himself that he worked by means of a code which he failed to see how he could intercept. However I will follow up his second proposition, set to work upon the construction of my model at once, and hope to be ready for the contest before the end of the month.[20]

Mr Phillips, reading Matthews' reply with 'some amusement', wrote to the *Daily Express* on May 9th 1910:

I am not aware that I have ever stated – at all events to the Press – that the model airship I am now exhibiting at the Hippo could not be deflected, directed, or controlled by another person. [Author's italics.] I thought I pointed out to Mr. Grindell Matthews very clearly when he saw me at the Hippo that there would be no difficulty in accomplishing the feat which was to form the basis of his 'challenge.' My only reason for replying to his letter as I did was to provide – as would undoubtedly have been the case – an amusing contest with my airship ... as a war vessel, my airship would carry instruments and apparatus which it is not possible to use on such a small model as is now exhibited, and with my code and other arrangements – which, for obvious reasons, I do not wish to publish – would make it practically impossible for any one to tamper with the 'electric waves' transmitted to the airship.

Matthews set about building his rival airship and financed by The Grindell Matthews Wireless Telephone Company he hired rooms in the New Passage Hotel, Pilning, on the banks of the River Severn in South Gloucestershire, and arranged for another workshop and radio station to be built in the grounds. During the winter months of 1910 he set up a small radio transmitter on the Black Rock in the River Severn estuary and was successfully transmitting radio broadcasts back to his workshop at the New Passage Hotel.

Along with the equipment installed at his New Passage workshop was a model Farman biplane and his motor launch, *Undine*. Not until now has the diversity of Matthews' work been fully appreciated as it is now apparent that Matthews was working on several different projects at that time. From the ceilings of one of the larger rooms in the hotel he hung the model airship with which he planned to challenge Mr Phillips. The airship was 21 feet long and between 5 and 6 feet at its widest point. With all the equipment attached to the airship's frame the whole thing weighed between 17 and 20 pounds.[21] Strangely, though, the outcome of the 'wireless duel' is not known: that is, if it ever took place.

Mr Phillips did patent his '*Improvements in or Connected with the Controlling of Aerial Vessels by Wave Transmitted Electricity*' on May 15th 1911. The device was a dirigible balloon that was remotely controlled by electromagnetic waves. He planned to build, and sell to the government at a cost of £100, an airship that was three times as big as the one he was demonstrating at the London Hippodrome and was 'capable of dropping dynamite sufficient to blow up the chief buildings of a large town'.[22] The patent shows how the framework was suspended from a balloon with forward propulsion provided by two propellers at the rear with a further two propellers providing upward thrust. Electric lamps illuminated when the propellers were in operation and an electromagnet could be remotely operated to release the tray allowing items to be dropped from the airship.

Matthews was elected to the Royal Institution in July 1910: he was just 30 years of age. Located in Albemarle Street, London, the Royal Institution was founded in 1799 and had a long list of distinguished members including such notable figures as Sir Humphrey Davy and Michael Faraday, whose Christmas lectures continue to this day. Marconi, another distinguished inventor, had been elected seven years before, back in 1903. The Royal Institution would have introduced Matthews to some of the leading scientists of the day and afforded him the opportunity to attend lectures and seminars enabling him to keep up to date with the latest goings on in the field of wireless technology. But not the type to rest on his laurels he planned to extend, further still, the range of his wireless communication from his radio station at Black Rock on the Welsh side of the River Severn at Portskewett, Monmouthsire.

On May 26th 1911 Matthews filed his third patent, GB191112730, '*Improvements in telephone Instruments.*'[23] This improvement enabled the distance of telephonic communication, without wires, to be increased by employing a more powerful electric current that was previously unattainable: 'The augmentation is effected by the provision of a plurality of contacting surfaces.'[24] By mid-1911 Matthews was achieving real and practical improvements in wireless communications technology. A closer look at the patents he filed shows just how productive he was at this time for in August 1911 another patent application was made: *Improvements in Wireless Telephony.*[25] The improvement was to the electronic transmitting circuitry where he made the addition of a high-frequency breaking-device. The higher the frequency of the radio signal emitted, the more precisely it can be focused and the smaller the footprint becomes. The footprint is the area cover by the radio antenna. The focusing of the signal on a smaller footprint can increase the energy of the signal and the smaller the footprint, the stronger the signal becomes and therefore the smaller the receiving antenna.

Jabez Wolffe, a well-known swimmer, had made several failed attempts to swim the English Channel during the early part of the twentieth century. In September 1911 he was planning to break the previous record set by another cross-channel swimmer and wrote to the press asking if they could put him in touch with Matthews. Jabez had the idea that Matthews could use his wireless telephone to report his progress as he swam the chilly waters from Calais to Dover. Always with an eye for a good bit of publicity, Matthews was impressed by Jabez and agreed to make the necessary arrangements for a live radio broadcast of the cross-channel swim. The plan was for Matthews to follow alongside Jabez in a tug called *Sophie*, where his Aerophone would be installed, and he would report his progress to a receiving station set up on the shore at Sangette, near Calais on the French side of the English Channel. In order to be able to make such a broadcast Matthews needed permission from the General Post Offices of both France and England, which they both readily gave, with the Secretary of the GPO in London Expressing 'great interest'.[26] The Admiralty had a wireless station at Dover and granted permission for the event to go ahead after Matthews gave reassurances 'that the test would in no way interfere with any wireless messages' and 'that it was as impossible for him to tap wireless messages as it was for the Admiralty to tap the Aerophone messages'.[27] Unfortunately however, due to bad weather, Jabez's attempt had to be postponed and finally abandoned. Between 1906 and 1911 Wolffe made at least twenty-two attempts to cross the channel but never succeeded. In one attempt made in 1911, he failed by yards and by less than a mile on three other separate occasions.[28] Matthews never got another chance to make his live cross-channel broadcast.

It was in that same month, September 1911, that Matthews attempted radio communication from the Railway Pier just outside the New Passage Hotel, to the Old Passage Beachley Ferry Inn, Gloucestershire, three miles upstream from the New Passage Hotel on the opposite side of the River Severn. Matthews was now making a serious attempt to see exactly how far he could transmit speech into the ether over Gloucestershire and the river Severn. Along with his team he set up a large canvas tent to house the radio transmitting equipment at Beachley. The radio waves produced by his apparatus were of very high frequency: so much so that there was no significant break between them, and consequently they were almost continuous in nature rather than the intermittent form of Hertzian waves that the Marconi system of wireless telegraphy was based upon. These high-frequency radio waves are much more faithful in transmitting and reproducing sound, meaning that the transmission of human speech could be continuously sustained and easily reproduced with all its fluctuations. This explained why Matthews' Aerophone was sensitive enough to broadcast the ticking sound of his pocket watch. Marconi was not able to do this; his system could not transmit voices, only Morse, the reason for this being that he wasn't able to generate high-frequency radio waves.

Matthews was using large antennae in an attempt to increase the range of transmission and sensitivity of reception. The antennae were attached to kites, one connected to the radio receiver and one to the transmitter, at a height of several hundred feet in the air. The problem with using kites is that they are at the mercy of the weather and very difficult to set up reliably 'in the afternoon the wind suddenly veered from the east to south and brought down the wire-laden kite which was connected to a battery on the shore, damaging it considerably'.[29] The *Daily Express* reported that

> it has become evident during the past few days that there has been an organized attempt to steal from Mr. Grindell-Matthews the secret of wireless telephony when yesterday morning Mr. Grindell-Matthews set off to the tent on the rocky coast of the Severn at Beachley, where his instruments are kept, his chief mechanic said 'I had to leave the tent for a while, and when I returned I discovered that it had been tampered with and one of the bolts had been cut clean off.' Matthews after making a close inspection noticed that 'wires had been crossed and muddled.'[30]

A few days earlier whilst setting up his apparatus 'one of his kites descended suddenly without any apparent cause, and he strongly suspects that one of the stays was cut so as to weaken it'.[31] Who was committing the alleged acts of sabotage? A rival commercial organisation? The government and Marconi were aware of his work and both would be implicated in the 'Marconi Scandal' that would break in less than a year's time. But despite technical setbacks and attempts of sabotage Matthews and his team of pioneering workers were having notable success.

Meanwhile descriptions of the Aerophone started to appear in the press, both in the UK and abroad, with one such account being printed in the *Black and White*, February 1910: 'Consisting of a small and portable set of electrical apparatus, with a receiver similar to that of a telephone ... messages could be heard distinctly and accurately at a distance of over seven miles.'[32] It was suggested that the Aerophone could be used for telephoning from an aeroplane or 'from a motor car whilst in motion'.[33] The *Bexhill-on-Sea Observer* reported on April 16th 1910 that 'the apparatus does not depend upon Hertzian waves ... no internal or high poles with earth connection are required, and the energy ... is directive'. The mobile phone had arrived in 1910. Finally, after much research and development and not inconsiderable cost, towards the end of 1911 the Grindell Matthews Wireless Telephone Company was planning to sell the Aerophone for £18 18s. Having perfected the portable apparatus, which had a transmitting range of a five mile radius, Matthews was confident that it would sell well, particularly when the equivalent Marconi sets cost £150. Had he been able to achieve this range in the previous year, 1910, when Sir Joseph Lyons was interested, he could well have collected the £500,000 that Sir Joseph had offered to the inventor. The Aerophone was far more elegant in its simplicity than Marconi's more cumbersome arrangement: a self- contained box containing everything required for the transmission of conversation with both transmitter and receiver being completely portable. The Marconi system was less advanced, requiring skilled operators, and capable of only transmitting Morse code.

'A WIRELESS TELEPHONE
Marvelous New Invention – The Aerophone
An Instrument Which Annihilates Space'

'Aerophones for £18 18s
New Invention to be put on Sale'[34]

'Anyone will be able to buy a complete outfit for Aerophoning for eighteen guineas in a little while,' said Sir William Bull, solicitor and Conservative MP, who became chairman of The Grindell Matthews Wireless Telephone Company after the death of Sir Clifton Robinson in 1910. He told reporters that 'this would probably be the price of the two necessary small boxes, and it is expected that before long the Aerophone would be in everyday use'.[35] It is worth noting here that, around this time, the government were planning to transfer the telephone network from private to public ownership, giving the government complete control over the telephone network. Although this wasn't to happen until December 28th 1911 when the Aerophone was going to be put on sale, the government were making plans for the transfer. The Aerophone would comprise the government's plans for that monopoly. In the summer months of that same year the 'Marconi Scandal' would erupt.

With all the publicity surrounding the Aerophone and with the plans they had for the nationalisation of the telephone network, it was inevitable that the government took an interest in Matthews and his Aerophone. Impressed with what the inventor claimed it could do, the Admiralty invited him to give a demonstration and so began a chain of events that would have serious and lasting consequences for both Matthews and the government. Eager to show the potential of the Aerophone, Matthews readily accepted the invitation. Because he had a further two patents pending relating to improvements he had made on the Aerophone he had requested that no technical experts be present and the government agreed to this. Matthews was happy that a journalist should be present at the demonstration and on September 29th 1911[36] he, along with his two assistants A.J. Alavoine and Mr Poncon, went to the Army Wireless Experiment Establishment at Aldershot and set up his equipment. Matthews was to hold a conversation between two distant rooms. Having set up the transmitter in the passage and the receiver in a room opposite the quadrangle, a distance of 50 yards, he was unable to establish a radio link and went to see if the problem lay in the other room. Having identified the problem Matthews returned to discover Lefroy, a government official, making detailed notes and drawings. An incensed Matthews ordered his assistants to stop the demonstration and immediately packed everything away. Witnessing all this, the journalist wrote an article in the *Daily Express* in which Matthews accused the War Office of taking advantage of the situation and trying to steal his secrets. An enquiry was held into the matter and confidential files in the National Archives make interesting reading. Lefroy was present at the enquiry and denies any wrongdoing. Admitting that in Matthews' absence he 'took pencil and paper and began to make notes for my report, so as to save time, whilst we were waiting for the return of Mr. Grindell Matthews'. Lefroy claimed that he 'did not open any apparatus or do anything other that that which was *natural for the purpose of writing a technical report*' (author's italics).[37] At the enquiry Lefroy said that, as far as he was concerned the Aerophone had no technical value. This incident raised two interesting points: why did Lefroy wait until he was alone before he started to take notes? And despite his low opinion of the Aerophone it is well documented that there had been numerous successful demonstrations of it. Matthews was to later deeply regret the allegations he made against the government and wrote an article to the *Daily Express*[38] claiming the whole incident was simply a misunderstanding. However this was to be the beginning of a relationship that was to be overshadowed with mistrust and suspicion that was to last for the rest of his career.

Matthews was still trying to increase the range and applications of his radio equipment and had heard about the work of W.R. Ferris who in 1910 transmitted wireless telegraphy, the transmission of signals, not speech, from an aeroplane.[39] Matthews was convinced that it would be possible to communicate with a pilot using his equipment and on Saturday, September 23rd 1911, a world first was made at Ely Racecourse in Cardiff when Matthews made radio communication with a moving aircraft. Bentfield Charles Hucks was a celebrated aircraft pilot who became famous as the first English pilot to fly upside down in a loop. Matthews invited Mr Hucks to take part in a demonstration of ground-to-air wireless communication. A Blackburn Mercury monoplane was fitted with the necessary apparatus with Hucks wearing a telephone receiver that had been modified so that it fitted to his ears and made it a hands-free device. A photograph of the occasion shows a smartly- dressed and youthful looking Matthews, with an Edwardian-style moustache, crouching over the controls of his radio transmitting apparatus as he attempts to make radio contact with Hucks. An initial test to establish radio contact was made with, firstly, the aircraft's engine turned off. Speaking to Hucks the test was successful, with Hucks being able to hear Matthews' voice. Matthews than asked for the aircraft's engine to be started for the second test. Again Hucks was able to hear Matthews speak to him, via the headset, over the sound of the roaring engine. The third and most exciting test could now be attempted. With the weather being agreeable and after lengthy consultation, the team decided to attempt radio communication with Hucks whilst in flight. Preparations and checks completed, Hucks slowly piloted his Mercury monoplane to an altitude of 700 feet. With Matthews and his assistants peering into the sky and seeing Hucks make his approach, Matthews spoke into the microphone of the throbbing radio equipment and whilst Hucks was flying over the aerodrome at 60 miles per hour Matthews was able to establish radio contact with him. Matthews felt a tremendous thrill, for here he was using the wireless equipment to speak to a pilot in flight! 'I am satisfied with the progress made, it is what I expected to prove. Mr. Hucks heard my voice distinctly.'[40] Matthews gave an interview at the Chepstow Hotel to the *Daily Chronicle*[41] where he said that he 'hoped to speak 40 miles shortly' and that he had 'spent 11 years developing his system and hoped to make his experiment in the following week'. It is interesting to note that he mentions that 'the government was interested'. He now moved his workshop from Pilning to London to prepare for further tests to increase the range of his wireless Aerophone.

All this work developing the Aerophone, setting up radio stations, building airships and giving demonstrations was enormously expensive and to date The Grindell Matthews Wireless Telephone Company hadn't made a penny in profit: it hadn't done any trading. On October 13th 1911 the company's annual general meeting became concerned at the soaring costs and details of financial problems came to light. The *Electrician* reported that the idea was to 'issue only a sufficient number of shares to provide necessary funds necessary to enable Mr. Grindell-Matthews and his skilled assistants to bring the wireless telephone to such a state of efficiency as would place the company in a position to sell its undertaking to a larger company at a profit to the shareholders'.[42] However 'owing to unexpected difficulties encountered by Mr. Grindell-Matthews' funds had become exhausted and shareholders were unwilling to provide any more. A major storm during the autumn of the previous year had flooded the workshop at Pilning, damaging equipment, which only added to the company's financial burdens. Matthews convinced the directors that the future of the Aerophone was bright and that all his work would bring them all financial dividends if only they would fund him for a little longer. The directors decided to place shares 'in other quarters' and sell property, valued at £16,575, in order to make more funds available and 'which enabled Mr. Grindell Matthews to continue his work'.[43] The directors were looking into the sale of the company's patents to France, America, Canada and Austria-Hungary. The company had, at that time, 20,000 shares but the directors 'were negotiating with a company to increase the balance of un-issued capital to £150,000 or £7.10s per share'.[44] So it wasn't all plain sailing but Matthews did have a financially viable company run by competent businessmen.

During 1911 Matthews corresponded with two of the leading science fiction writers of the time. Rider Haggard was a prolific writer of science fiction with *King Solomon's Mines*, published in 1885, being one of his best-remembered works. Matthews was a keen fan and read his works from an early age and first came across the word 'Aerophone' when he had read the novel *Stella Fregelius*, published in 1903. In the novel Haggard vividly describes a device called the Aerophone which was a 'wireless telephone, made in pairs' that didn't require 'high poles or balloons' invented by Morris Monk who had worked on the device for 10 years.[45] Using his Aerophone Morris speaks to Stella, the heroine 'although separated by a raging sea'.[46] Rider Haggard wrote to Matthews calling his attention to his works and asked if they could meet, which they did. They discussed their mutual interest in science fiction and in particular the inventor's Aerophone, how it worked and the future possibilities of wireless technology. It pleased Matthews no end to be able to name one of the leading writers of science fiction as one of his friends. The second author Matthews corresponded with was Marie Corelli who was another prolific author of science fiction. Her work was treated with widespread criticism in the press, to which she was highly sensitive, and she lived a relatively reclusive existence in Stratford-upon-Avon. An excellent account of her life is given in *The Mysterious Miss Marie Corelli*[47]. *Life Everlasting*, a mixture of science fiction and fantasy was published in 1911 and, after reading her work, Matthews wrote to Miss Corelli telling of his admiration of the book and to invite her to give a public speech about his Aerophone. She replied to him at the Chepstow Hotel saying 'your work is profoundly interesting to me'[48] and thanking him for his invitation but felt, because of her curious relationship with the British press, who she said were 'united as they are in an effort to ignore and ridicule me',[49] she had to decline. Writing to Matthews she said 'I shall watch all your movements with eagerness.'[50] Miss Corelli asked Matthews for a signed photograph and whether he could call in at her home, 'Mason Croft' in Stratford-upon-Avon. This was the only time the two met and they discussed the Dream Yacht that appears in *Life Everlasting*. The Dream Yacht is a vessel whereby 'sails are our only motive power, but we do not need the wind to fill them. We can generate a form of electricity from the air and water as we move.'[51] They had a long chat about their respective views and the possibilities of science and technology.

Having now perfected the Aerophone Matthews applied for a further two patents: *'Improvements in Wireless Telephony'* (1912)[52] and *'Improvements in Means for Producing Electromagnetic Waves of High Group Frequency and in the Application Thereof to Wireless Telephony'* (1913).[53] This details the way in which Matthews was generating a carrier wave to transmit speech by wireless and 'whereby the quality of speech transmission may be maintained over lengthy periods'. Matthews had given his address as 'The Royal Societies Club, 63 St. James' Street London'. The Royal Society Club was founded in 1894 and was a forum for many distinguished and learned people and societies. It gave Matthews excellent opportunities for networking and making valuable contacts in the scientific community. Matthews was no scientific fraudster: he was in touch with others in the same field, willing and eager to collaborate. The patent goes on to give details of producing electric oscillations that have a wave frequency exceeding the audible limit produced by a spark-gap. The device also had a microphone which enabled voice to be imposed on the oscillations. This is basically giving details of a device that can generate waves that are much better at transmitting human voices by wireless, something that, at the time, Marconi was struggling to achieve. Marconi wouldn't be able to generate and use a reliable carrier wave for voice transmission until 1920 when he was working at an experimental station in Writtle, Essex.

At the time Matthews was working with a William Ditcham who was the Chief Engineer for The Grindell Matthews Wireless Telephone Company and they were to work closely together during the pre-war years. Ditcham, with distinguished credentials himself was, along with numerous other inventors in this field including De Forest and A.A. Swinton, who was President of the Wireless Society of London, a member of the Institute of Radio Engineers. Presided over by Louis W. Austin, the organisation published many technical papers on wireless communication and brought together like-minded inventors and engineers with a membership that extended to the United States. What is very interesting about Ditcham, though, is that during the 1930s he went to work for Marconi's Wireless Telegraph Company in London. Whilst working for Marconi he made numerous patent applications, thirty in total, both in England and Canada, relating to improvements made to carrier wave and radio transmitters. This shows his knowledge and expertise in carrier waves that are essential for the wireless transmission of speech.

It is interesting to note that the 1913 patent filed by Ditcham and Matthews has been officially marked 'void: sealing fee not paid'. The sealing the fee is the amount payable, by the applicant, in order for the patent to be granted. This may be symptomatic of the financial problems, reported in the previous October, that The Grindell Matthews Wireless Telephone Company was having at that time. The 1912 patent was an improvement to telephones that allows for 'a more powerful electric current to be effectively employed that has hitherto been possible, whereby telephonic communication through space unconducted by wire may be extended to an increased distance'. Matthews was striving to increase the distance at which he could transmit speech, for he saw that this would supersede Morse as the most effective way to communicate.

Late one blustery night, in January 1912, a sailor was crouched over a wireless set in the tiny communications cabin of a large fishing trawler making its stomach-churning way back across the North Sea to Norway. Due to the rough sea he was having some difficulty tuning in the controls of the wireless Magnetic Band Detector, a device for transmitting Morse code. Eventually, after tuning the apparatus, he wearily began pressing the Morse key and tapped out the message the skipper had asked him to send back to headquarters. Sitting back he waited for a reply, with his ears getting hot and sticky under the uncomfortable ear phones when to his utter amazement instead of the familiar monotonous sound of dots and dashes he heard a voice: 'Can you hear me? I'm going to play some music. Can you hear it?' And so on January 31st 1912 'The Voice of the North Sea Ghost' entered the news with headlines like 'Mysterious Voice! Puzzled Wireless Experts!'[54] appearing in newspapers. Stories describing how sailors, when listening in on their wireless sets, heard what they thought was a human voice appeared in the newspapers:

> Has a practical solution to the difficult problem of carrying the human voice through space by wireless been found? Marked success has clearly attended the efforts of someone in this direction, but who he is, and from what part of the world his operations are being carried on, at present remains a mystery. A day or two ago human speech was 'picked up' by several independent experts who have installations for practicing wireless telegraphy in this country. So far as is known, human speech has never been picked up in this way before. Hitherto speech has only been picked up by persons working in collusion with the transmitter, and with apparatus specially designed for the work. The fact that this message or a series of messages in actual speech has been received by various wireless apparatus which, in the opinion of a celebrated wireless expert, have hitherto been regarded as entirely unsuitable marks a very definite forward step in the solution of the wireless problem. Who is he?[55]

This all generated speculation, debate and interest in the press and scientific community. The real possibility of a new technology was arriving. Wireless telephony was in its infancy, still a stumbling, fledgling technology but the 'North Sea Ghost' was a Rubicon, a definite leap forward. 'The Voice of the North Sea Ghost' was voicing progress, and yet this leap was being made by an unknown! An unnamed company identified with wireless telegraphy claimed to be the 'voice' but at this point Matthews came forward and met with the press to declare that he was 'The Voice of the North Sea Ghost'. His well-publicised work detailing his successful attempt at wireless communication with an airman in flight, back in September 1911, gave him sufficient credibility. But the press, sensing a scoop, called for a demonstration to an audience of credible witnesses and experts. Mr Ross, an amateur wireless enthusiast and a member of the Wireless Club, and one of the radio hams who first heard Matthews whilst listening on his wireless radio set at his home in Dalston, London, issued Matthews with a challenge. He wrote to the *Daily Express*:

If Mr. Grindell Matthews is convinced that the messages I received came from his experimental station in Gloucestershire, I challenge him to undertake a series of tests with me under the observation of competent judges, so that the mystery may be laid to rest once and for all. I suggest that the tests should consist of messages sent off from Mr. Matthews' instrument at Black Rock at an arranged hour, so that I can be ready to receive them on my machine at Dalston.[56]

Never one to shy away from a challenge Matthews telephoned the *Daily Express* two days later to accept and all the necessary arrangements were made. The editor of *Daily Express* sent Matthews a telegram on the morning of February 5th which read 'Ross listening tonight. Please wire me now test speech numbers you're phoning for verification. Will you also play well-known tune, wiring name.'[57] With a journalist, to act as a witness, sitting with Mr Ross, they both settled at the apparatus set up in the corner of Ross's kitchen at 7 o'clock to tune in the radio receiver and scan the airwaves. As arranged, Matthews was due to transmit at 8 o'clock. Just before the hour was reached Ross tuned in and picked up the sound of a carrier wave. After a few moments of tense expectation came, into the Dalston kitchen, the words: 'Ross. Ross.' Ross recognised the voice and made some fine adjustments to his receiver and then he heard more clearly: 'Hello. Matthews here. Matthews here. Hello, Ross.'[58] The weather, which was stormy that night, was making the signal quite weak but Ross made further adjustments with the dials and succeeded in getting still yet a better signal. 'Hello Ross. Matthews here. Can you hear me? One, two, three, four, five, six, seven, eight, nine, ten. Ninety-nine. Have you got it, Ross? Can the *Daily Express* reporter hear me? I'm going to play "Two Eyes of Grey". I hope you can hear it. Matthews calling Ross.'[59] The distance between Black Rock on the River Severn and Dalston, in Hackney, London is 110 miles.

Earlier that same year Matthews had been working with Mr Herbert Henke, one of his assistants, to fix an aerial on the roof of the *Western Mail* buildings in Cardiff. The aerial was 40ft in height and shaped liked an inverted L. This was then connected to radio receiving equipment set up in the directors' boardroom.

A transmitting aerial had been erected on the roof of the Westgate Hotel in Newport. On January 20th 1912 the *Civil Service Gazette* reported that 'Newport and Cardiff are now linked together by wireless telegraphy and wireless telephony. Wireless telegraphy works by means of a code as in ordinary telegraphy, whereas the transmission of the human voice by means of air-waves of high frequency marks a distinct advance on the Marconi system.' Work progressed for about four weeks until finally on February 27th 1912 Matthews transmitted the first ever wireless press message from Newport to the *Western Mail* buildings in Cardiff. The message, consisting of 130 words, was to the directors thanking them for their assistance. The message was printed in the *Western Mail*:

> I take this opportunity of thanking the directors of the *Western Mail* for the facilities they have afforded me in my experiments between Cardiff and Newport. This is the first message ever to be transmitted by wireless telephone to any newspaper, and I am glad that the unique distinction, which will be of great historical value in the future, has fallen to the lot of the *Western Mail*, whose proprietors have met my convenience so handsomely.
>
> When the Aerophone has been adopted for commercial purposes, and its application to naval and army purposes has been accepted by the British Government, Cardiff and Newport, and the *Western Mail* in particular, owing to their association with me in these successful experiments, will all share fully in the glory of the achievement.[60]

Later in March, 1912, Matthews bought two Schneider motor cars and fitted them with transmitting and receiving apparatus and travelled around giving demonstrations of wireless broadcasting. The *Bournemouth Guardian* reported that

> interesting wireless telephone experiments have been carried out in the west of England by Grindell Matthews. The experiments were conducted from motor-cars which were furnished with telescopic masts, and conversations were held between the parties occupying the cars, which were situated as far apart as Newport and Bath (26 miles distant) and Newport and Warminster, a distance of 50 miles.[61]

In July 1912 Matthews was invited by King George V to demonstrate his wireless apparatus at Buckingham Palace, an event that was reported in *The Times* Court Circular.[62] During the same month Matthews gave a demonstration of his Aerophone to Lloyd George, the Chancellor of the Exchequer and Colonel Seely, Secretary of State for War. The Regent Carriage Company in Fulham was the venue and Matthews used the same cars that he used for the Royal demonstration. Lloyd George was seated in one car with Col. Seely in the second car. They were able to hold a conversation with each other and the *Daily Express* commented on July 26th 1912:

> Although their visit was a private one, there is little doubt that they tested the wireless telephone with a view to considering its value for Government use. Mr. Grindell Matthews was invited to give a demonstration at Aldershot at an early date. What struck Col. Seely most was the value of the Aerophone in conjunction with the monoplane in time of war.
>
> He declared that it would make the monoplane of immense value for scouting purposes, as the aviator could set out across the country giving information as he went by wireless telephone of the disposition of the enemy's forces. It was hoped that Mr. Winston Churchill would be present, but he was prevented by other engagements from joining the party. The instruments are portable, and one of the chief features of the wireless telephone is the fact that you can recognize the voice of the person speaking.

Thoroughly impressed, Lloyd George gave Matthews an assurance that he would do all that he could to help him further the cause of his research: an assurance, however, that would later be seen to waver.

It was around this time that 'an invitation came from France' for him to demonstrate, with his wireless equipment, a cross-channel conversation between Folkestone and Boulogne.[63] Matthews accepted the offer with eagerness and set up radio stations at the Grand Hotel in Folkestone and the Folkestone Hotel in Boulogne. With all the necessary preparations completed and the radio installations in place Matthews, to his utter astonishment, was told by the Postmaster General that his application for a licence permitting him to transmit across the

Channel had been refused on the grounds that his wireless signals 'might interfere with the wireless telegraph' used by the Post Office.[64] Because of this Matthews was unable to press on with his planned cross-channel wireless conversation and had to dismantle all his equipment. The sinister goings-on that led to the refusal became apparent in an incident referred to as the 'Marconi Scandal'. During the summer months of 1912 allegations were made about parliamentary members of the Liberal government who, it was alleged, had used information about government plans for the Marconi Company for personal profit. In the previous year Prime Minister Asquith gave the go-ahead for the construction of state-owned wireless stations to be built throughout the British Empire. At the request of the Prime Minister, the Postmaster General, Herbert Samuel, was given the task of finding a company to build the overseas wireless stations. Samuel chose to give the work and subsequent contracts to The Marconi Wireless Telegraph Company, the chairman of which happened to be Godfrey Isaacs, a close friend of Herbert Samuel. Isaacs' brother, Sir Rufus Isaacs, was the Attorney General for the Liberal government. Consequently, Marconi's shares rose enormously. Hilaire Belloc, the editor of the political journal *The Eye-Witness*, had suspicions that members of Asquith's government had, using inside political information, been buying shares in the Marconi Company, knowing that they would be awarded the contract. Belloc printed his suspicions in *The Eye-Witness* making allegations that Sir Rufus had made £160,000 from shares he had bought.

A parliamentary inquiry was held, in 1913, into the allegations made against Asquith's government. The inquiry found that Godfrey Isaacs had bought £10,000 worth of Marconi shares, and his brother Sir Rufus had bought 10,000 £2 shares and had sold 1,000 of them on to the Prime Minister. The inquiry also found that Herbert Samuel had bought 3,000 shares as had the Liberal Chief Whip. Although the inquiry did find Lloyd George, Herbert Samuel and Sir Rufus had profited directly from the policies of the government they were found not guilty of corruption.[65] It's no surprise then that the British Government refused Matthews a licence to transmit radio signals: after all, he was a rival to Marconi, and a serious one at that. But this does prove that the government did see Matthews as a serious contender in wireless technology. Matthews issued a challenge to the Marconi Company to test their respective wireless equipment. However Godfrey Isaacs, their Managing Director, declined saying that

> with regard to any challenge that may be thrown out to the Marconi Wireless Telegraph Company I intend taking no notice of it whatever. When the thing is practical, commercially, working success, it will announce itself. Until that time no public demonstration, of it, or announcement about it will be made.[66]

During the latter part of 1912 Matthews, with finance from The Grindell Matthews Wireless Telephone Company, set up two radio broadcasting stations: one in Northampton and one at Letchworth in Hertfordshire. Three of the first transmitting valves manufactured by Cossor in England were made to Matthews' specifications and used at his Letchworth radio station. Cossor, a long-established manufacturer of electrical components including cathode ray tubes for television receivers, made radio sets in kit form for the burgeoning amateur wireless market. During the Second World War the company's technicians worked on radar defence systems.

As the storm clouds of the First World War were gathering on the horizon over Europe, Matthews made a quantum leap in his work as he managed to transmit speech over a distance of 670 miles between two French villages, Sangatte, just outside Calais, and Barraques in the South of France. The transmitting station was set up on the sands at Sangatte, and the receiving station on the ground floor of the Hotel de la Plage in Barraques where Mr Alavoine, Matthews' assistant, was based. At each station kites were used to hold aloft the aerials. A journalist for the *Daily Express* reported that when the radio equipment was operating 'a long whooing sound throbs in the room, sending vibrating echoes from wall to wall'. The 'buzzing of dynamos and the startling flickering of electric sparks' were seen by startled guests in the hotel.[67] But it was all worthwhile for Matthews established a radio link with the journalist excitedly reporting that 'the voices carried across the wind-swept sands, were audible above the noise of the waves breaking on the shore'.[68] His success, widely reported in the press, brought him to the attention of the French Government who made enquiries after the inventor.

Matthews' quest to extend the range over which he could transmit speech was obsessive, for between the end of 1909 and the middle of 1913 he filed no less than seven patents relating to improvements in wireless telephony. Three patents appeared in just one year, all of them refining his Aerophone device so that it could transmit speech more faithfully over ever greater distances. *'Improvements in Arrangements for Producing Electro-magnetic Oscillations particularly for use in Radio Telephony'* appeared in June 1913, stating that, 'Electro-magnetic waves of high group frequency may be radiated, which have a definite wave length, small dampening, and *are of great regularity and constant intensity and very suitable for use in radio telephone*'[69] (author's italics). Another patent[70] detailed the design of switching gear he and Ditcham had added to the Aerophone, whereby a switch could be used to alternate between the receiving circuit and the transmitting circuit. One unit could act as both transmitter and receiver, a truly remarkable technological step forward. Matthews was an able and diverse electrical engineer, which is only too apparent in the number of patents he filed before the outbreak of the First World War: a total of nine by the time he was thirty-three. But he didn't just confine his inventive mind to wireless communication. He had already invented an automatic piloting device back in 1909 and in 1913, the last patent he filed before the war was for a wireless burglar alarm.[71]

In January 1914, just as Matthews was planning his most ambitious radio transmission to send speech from Newfoundland, Canada across the Atlantic, news reached him that his creditors had presented a petition to the Bankruptcy Court and bankruptcy proceedings were started against The Grindell Matthews Wireless Telephone Company Ltd. The research and development of the Aerophone had been enormously expensive and Matthews, never happy just to concentrate on one experiment, had been ordering vast amounts of equipment, using craftsmen and employing technicians for some considerable time. The bankruptcy court recognised that the failure of the company was attributed to the fact that the completion of the work to perfect the Aerophone took much longer and therefore incurred greater expense than was originally contemplated or provided for. By the end of 1913 the company had 142 share holders totaling £10,562 in subscribed capital. The assets of the company including the radio stations at Northampton, Letchworth, London Wall Buildings and at 8 High Street in Highgate, amounted to £31,000. But during the intensive research and experiments bills had been left unpaid for months and the company's creditors, of which there were 43, were getting impatient. Directors' fees and monies for electrical equipment and services were all outstanding. Much of the equipment, often manufactured to Matthews' own specifications, such as dynamos, generators and aerials installed at the Letchworth station, was on credit. And not forgetting the outstanding insurance bill for the Aerophone. It was all so incredibly expensive and the creditors, not unreasonably, wanted their money. Matthews has been accused of spending his investors' money on high living and expensive hotels. True, he did often stay at hotels but he used them as offices, it simply made practical sense. He did have the use of a chauffeur-driven car as well but when one examines the list of creditors it is easy to see that he was no mere crank simply in the business for easy money and high living.

But Matthews remained undaunted and despite the mounting financial problems, totally dedicated to his work he went to Paris with the London editor of *Le Journal* to make all the necessary arrangements to make a radio transmission from the Eiffel Tower to a French warship in the Mediterranean. Just a few years earlier in 1908 Lee de Forest installed a wireless transmitter on the top of the Eiffel Tower for broadcasting music played on a gramophone which was heard up to a distance of 500 miles.[72] At this time news also reached him that the equipment for his planned broadcast across the Atlantic was ready and waiting to be sent to Canada. But the bankruptcy proceedings against him weren't his only problem when on June 28th 1914 the opening chapter of the First World War was written with the shooting of Archduke Franz Ferdinand, the heir to the Austro-Hungarian throne, and his wife in Sarajevo by Gavrilo Princip. Matthews received a telegram from the Postmaster General: *Return at once. Dismantle wireless station in Letchworth and Northampton.*[73] Reaching Letchworth four days later, amidst the panic that was sweeping Europe, Matthews was alarmed to discover that Post Office engineers had already dismantled the station's transmitting aerial, locked the buildings and placed a guard on duty, preventing him from getting access to his precious equipment.

Returning immediately to London he discovered his business manager, Hon. Nelson Hood, had left for active service in France. News of his tragic death whilst fighting in France would reach Matthews only weeks later. Worried about his equipment languishing at Letchworth, Matthews wrote a letter to Lloyd George asking to be allowed access to his radio-telephone equipment. The reply he received simply referred him to the Post Office. After a lengthy exchange of correspondence Matthews was eventually given permission to retrieve his equipment, after a period of two years. Upon arrival at Letchworth his worst fears were realised for all the buildings were neglected and crumbling: the damp had ruined his electrical equipment and it was beyond repair. Matthews was given the derisory sum of £10 10s[74] compensation by the government and decided, under the circumstances, to cut his losses, for Europe was still engulfed in a World War.

Notes

1. http://gb.espacenet.com/ patent number GB190925639 and GB190928083 (accessed 04/05/07).
2. Ibid.
3. Ibid.
4. *Daily Express*, May 7th 1910.
5. Ferry, Georgina, *A Computer called LEO: Lyons Tea Shops and the World's First Office Computer* (Fourth Estate, 2003).
6. Barwell, p. 20.
7. Ibid.
8. Ibid.
9. National Archives J/13/6587.
10. National Archives J/13/6597.
11. www.gtj.org.uk/en/blowup1/28822 (accessed 22/04/07).
12. Barwell, p. 20.
13. Ibid.
14. Ibid., p. 23.

15. *Daily Express*, May 3rd 1910. Swansea County Archive D/DZ 346/1.
16. Ibid., May 4th 1910. Swansea County Archive D/DZ 346/1.
17. Ibid.
18. Ibid., May 6th 1910.
19. Ibid., May 4th 1910. Swansea County Archive D/DZ 346/1.
20. Ibid., May 6th 1910.
21. *Western Daily Express*, June 6th 1910.
22. *Daily Express*, May 4th 1910. Swansea County Archive D/DZ 346/1.
23. http://gb.espacenet.com/ (accessed 04/05/07).
24. Ibid.
25. Ibid.
26. *Daily Express*, September 16th 1911. Swansea County Archive D/DZ 346/5.
27. www.dover.gov.uk/museum/focus/focus4.asp (accessed 22/02/07).
28. Ibid.
29. *Daily Express*, September 15th 1911. Swansea County Archive D/DZ 346/5.
30. Ibid.
31. Ibid.
32. *Black and White*, February 13th 1910.
33. Ibid.
34. Ibid.
35. Ibid.
36. National Archives ADM 116/4766.
37. Ibid.
38. *Daily Express*, January 1st 1912.
39. Sarkar, p. 103.
40. Barwell, p. 31.
41. *Daily Chronicle*, October 1911. Swansea County Archive D/DZ 346/5.
42. *Electrician*, October 20th 1911. Swansea County Archive D/DZ 346/5.
43. Ibid.
44. Ibid.
45. Barwell, p. 25.
46. *Western Mail*, February 2nd 1912.
47. Ransom, Teresa, *The Mysterious Miss Marie Corelli. Queen of Victorian Bestsellers* (Sutton Publishing, 1999).
48. Barwell, p. 26.
49. Ibid.
50. Ibid., p. 27.
51. Ibid., p. 26.
52. http://gb.espacet.com/ patent number GB191112730 (accessed 04/05/07).
53. http://gb.espacet.com/ patent number GB191206486 (accessed 04/05/07).

54. *Western Mail*, February 2nd 1912.
55. Barwell, p. 34.
56. Ibid., p. 35.
57. Ibid., p. 36.
58. Ibid., p. 37.
59. Ibid.
60. *Western Mail*, February 27th 1912.
61. *Bournemouth Guardian*, March 26th 1912.
62. *The Times*, July 5th 1912.
63. Barwell, p. 41.
64. *Daily Express*. Swansea County Archive D/DZ 346/5.
65. www.spartacus.schoolnet.co.uk/PRmarconi.htm (accessed 28/10/07).
66. *New Zealand Herald*, May 23rd 1924. Swansea County Archive D/DZ 346/1.
67. *Daily Express*. Swansea County Archive D/DZ 346/5.
68. Ibid.
69. http://gb.espacenet.com/ patent number GB191206486 (accessed 04/05/07).
70. http://gb.espacenet.com/ patent number GB191312157 (accessed 04/05/07).
71. http://gb.espacenet.com/ patent number GB191313065 (accessed 04/05/07).
72. Sarkar, p. 102.
73. Barwell, p. 42.
74. *News Review*, March 11th 1943.

CHAPTER 3

Into the Light – Dawn

With the world now at war, Matthews recalled his earlier experiments with the wireless-controlled torpedo and airship and considered the possibility of developing the technology further still. Both the torpedo and model airship were remotely controlled with radio waves but he thought of another method of control, a beam of light. Matthews' pioneering use of selenium in electrical circuits to remotely control various devices truly places him amongst the great inventors of the twentieth century. A beam of light can send rapid frequency pulses, 'flickering' in effect and, thought Matthews, just as with different radio frequencies, could be used to influence an electrical circuit via a selenium component. Selenium is a non-metallic element discovered, in 1817, by a Swedish chemist called Jöns Jakob Berzelius. The word selenium is derived from the Greek 'selene' meaning 'moon' and the element has the marvellous property of converting light directly into electricity. Its electrical conductivity varies with the intensity of light: the brighter the light the greater its electrical conductivity. It is because of this property that selenium is widely used in photoelectric devices and solar panels.

After some initial experiments carried out at his workshop at The New Passage Hotel, Matthews went to Edgbaston, in Birmingham, to continue his experiments with Mr B.J. Lynes, who was later to become his chief technician and would remain with Matthews for many years. Matthews had heard of Bernard Lynes, an electrical engineer, 'who had built up an enviable reputation in research work'[1] and invited him to join his research team that was building a small boat that could be controlled with a searchlight beam. What was uniquely special about the boat Matthews and his team were constructing was that it used a relay apparatus and a selenium bridge as a control unit or 'pilot' to control its movement. Matthews hoped that, once perfected, the 'pilot' device could be installed in various devices so they could be controlled remotely by a searchlight. He went to call on Dr Edmund Edward Fournier D'Albe, a scientist and philosopher, who was a renowned authority on selenium and author of *The Moon Element*, an extensive account on the uses of selenium. Born in 1886, D'Albe is widely remembered as the inventor of the Optophone, a device that allowed blind people to 'read' print by scanning ordinary printed characters and producing a musical note that is characteristic to each word or letter thereby enabling a blind person to 'read'. Matthews had asked E.E. Fournier D'Albe to help with the design and testing of the selenium bridge that was part of the 'pilot' device and 'consisted of an octagonal cylinder composed of eight graphite selenium tablets carefully selected for their uniformity of quality'.[2] This arrangement enabled the boat to be controlled from any direction. The selenium bridge that acted as a receiver of the impulses emitted by the searchlight was connected to a revolving commuter. This could be made to turn by successive flashes from the searchlight shining on the selenium in the relay device. At each turn the commuter would act upon a different switch, completing an electrical circuit, which brought about a corresponding action. It was possible to select which switch to function by adjusting the frequency pulses of the searchlight.

Experiments were carried out in relative secrecy on the Birmingham waterworks reservoir. The boathouse there was turned into a testing workshop and equipment store which housed the generator that was used to power the searchlight. 'The reservoir was specially guarded night and day by police, whose vigilance kept away intruders.'[3] However, after working there for a few days Matthews and his assistants became aware of a 'fisherman' who regularly pitched up at the reservoir in pursuit of his solitary pastime. Becoming a little bit suspicious of his regular appearances the local police made inquiries. Having spoken to him the authorities discovered that he was a Belgian refugee and were satisfied, although not completely, that he was genuine and simply out for a good days fishing. Still feeling a little uneasy, Matthews' technical staff made sure that there was scant opportunity for the Belgian to see what they were doing. Months later Matthews was to discover that the British Government was offered, at a price tag of £3,000,000, a device similar to the one he had been working on in Birmingham. This offer had been made by a Belgian inventor who was living in Birmingham and was the same person who was keen on fishing during those early trials at Edgbaston.

After several weeks of laborious testing, finally the remotely controlled boat was ready for a complete test run. In the early hours of the morning the Birmingham waterworks reservoir witnessed some unusual activity as Matthews and his team prepared everything for that first trial. With light spilling out from the tiny workshop onto the still surface of the water, a confident Matthews, standing in the receding darkness, switched on the searchlight and focused it onto the turret of a sleek boat, 16 feet in length, which lay motionless in the still water. With the searchlight carefully aimed at the vessel it slipped quietly from its mooring and headed across the water. The searchlight used for the test had a 24-inch Parsons mirror attached to it and with a small adjustment of the console the frequency of the pulses would change, causing the revolving commuter to switch circuit, and the vessel to alter its course. With Lynes directing the searchlight and Matthews at the control panel they were in complete control of the vessel as, between them, they manoeuvered it across the surface of the still reservoir. Their excitement grew: with the staff crossing their fingers and eagerly looking on, the vessel sailed out across the reservoir, leaving a gentle wake, performed various manoeuvres with swift ease and finally darted between two marker buoys. The method of using the searchlight beam to control the vessel seemed wholly reliable. With the first complete trial finished the searchlight guided the vessel back to her mooring just as dawn was breaking and so Matthews christened the boat '*Dawn*'.

So impressed was Commander Charles Craven, RN, one of the onlookers invited by Matthews to watch the first test run, that he immediately contacted Lord Fisher, chairman of the Board of

Invention & Research, and told him about what he had just witnessed. The day after the test Matthews received a telegram from Lord Fisher asking him to meet him at his home in Dungavel, Scotland, to discuss his work with *Dawn*. Accepting the invitation, Matthews made the long overnight journey by train, arriving in the morning. Years later Matthews recalled that first meeting and said of Lord Fisher 'from that stocky, well-knit frame and bulldog face radiated a force and energy more dominant and holding than from any man I have ever met'.[4] Lord Fisher was quick to realise Matthews' ability as an inventor and their mutual respect for each other grew over the course of the next few years. 'One of Lord Fisher's proudest possessions [was] a new kind of burglar alarm' invented by Matthews.[5] The gadget, housed in a small, neat, wooden box sat on Lord Fisher's desk. When a light was turned on in the room an alarm bell would ring. The same effect was caused if a flashlight was shone upon the instrument. Matthews had a great understanding of the nature of selenium and its applications and it's worth relating here a story told by Barwell. Whilst in his London flat, one evening, an idea occurred to him; was it possible to use the light reflected from Venus, nearly 24 million miles away, to 'act upon selenium and ring a bell?'[6] So one evening, in his London flat, he set up a small electrical circuit comprising of electrical wiring, an electrical bell and a small selenium bridge. As the evening light fell above the noise of the London traffic he placed a selenium bridge over the eyepiece of his telescope, which he very carefully focused on the celestial path of the planet Venus, and waited. As Venus slowly sailed across the heavens and moved onto the very spot where the telescope was focused, the bell rang.

Lord Fisher had built up a formidable reputation as a maverick leader and served on the Board of Invention & Research after his dramatic resignation, as First Sea Lord of the Admiralty in May 1915 over strategic differences with Winston Churchill regarding the battle of Gallipoli and the planned invasion of Turkey. Churchill later resigned as first Lord of the Admiralty following the failure of the campaign. Fisher was an unconventional high-flyer with a distinguished service record and a reputation for readily dismissing incompetent officers under his command. He worked hard to modernise the British Fleet, endeavouring to make it more efficient by encouraging the use of larger battleships, and was a strong advocate of submarines, an interest that would again bring him into contact with Matthews later on. After a distinguished military career Lord Fisher died in 1920 and was buried with full military honours during a state funeral at Westminster Abbey.

During their first meeting it soon emerged that they both had a mutual distrust of the Admiralty, and Matthews realised that he had found a friend whom he 'could turn to in any doubt or difficulty or delay in anything connected with the work in hand'.[7] Their discussion centered on the technology that was the basis of *Dawn*, its development and possible wartime applications. After Matthews explained the workings of the pilot device and how versatile it could be, Lord Fisher said he would arrange a demonstration of *Dawn* and asked him if it was possible to develop the technology as a potential solution to the threat posed by German Zeppelins. Zeppelins, named after Ferdinand von Zeppelin, were used to great effect during the First World War. With a huge envelope, shaped like a cigar, filled with extremely flammable hydrogen gas, powered by three engines, with two gondolas fixed underneath the rigid frame, they were mainly used for reconnaissance and could reach high altitudes in a short time. Although they were able to carry large loads they were slow-moving and therefore vulnerable to anti-aircraft fire and bad weather.[8] Matthews saw 'no reason why aerial torpedoes should not be directed in the air with the same ease as "*Dawn*" was controlled in the water'[9] and thought it would be possible to install the 'pilot' into an aerial torpedo so that its flight through the air could be controlled by a searchlight beam. E.E. Fournier D'Albe in his book *The Moon Element* recorded that Matthews 'rashly offered to construct and control an aerial torpedo boat to be steered automatically from the ground and to attack the enemy in mid air'.[10] After his meeting with Lord Fisher, Matthews returned to London later that day full of optimism and confident that he had gained an intelligent and influential ally.

Keen to develop an aerial torpedo, Lord Fisher, Commander Usborne, Commander Craven and Matthews met up at an address in Berkeley Square, London and were 'entrusted with the task of bringing into practical application an invention so fraught with immense possibilities that it will certainly if successful produce incalculable results as regards the termination of war'.[11] Lord Fisher expressed a need that 'stringent secrecy be maintained and the utmost celerity employed to bring the invention into action against the enemy'.[12] Craven was given the task of arranging all the necessary materials and facilities to design, plan and construct the aerial torpedo. With his earlier work on the radio-controlled airship constructed for the wireless duel back in 1910, Matthews was well placed to discuss the practicalities of such a venture and he convinced Usborne and Craven that he could direct and explode, with the required accuracy, an aerial torpedo. Matthews left the meeting in high spirits, convinced that he was dealing with men who were genuinely eager to press on with the project for the benefit of the nation's defence during a time of intense international conflict. This was much to Matthews' relief after his previous experience of government bureaucrats and Whitehall red tape.

The day after their meeting Matthews went, along with Commander Usborne, to King's North, near Chatham, to inspect the small type of airships known as 'blimps'. Blimps, otherwise known as non-rigid airships, were used extensively during the First World War for coastal protection against the threat of ships and submarines. They were simply a huge gas bag with a gondola, or fuselage, fixed underneath. An engine attached to the fuselage would propel the 'blimp' forward. Inspecting the 'blimps' closely, Matthews could see both the pros and cons of using them, but he thought that better protection could be achieved with the use of a 'torpedo-shaped gas bag fitted with a rudder' that could be 'controlled by a searchlight' using the selenium pilot, that had been so effectively demonstrated with *Dawn*.[13] What Matthews had in mind was a more sophisticated craft, called a dirigible, based on his radio-controlled airship design, but with greater flexibility than a conventional 'blimp'. A dirigible is a rigid airship able to carry a much larger fuselage than a 'blimp' and what Matthews envisaged was an aerial craft that could be guided, release bombs and also detonate them in mid-air, all under the influence of a searchlight. Work on the project was started immediately with early trials offering promising results – only to be followed by a fatal accident that was to end the whole project.

At the time when Matthews began his association with Lord Fisher, Commander Usborne, who was to work closely with Matthews and was in command of a government experimental and research facility during the development of the aerial torpedo, had thought of another method for destroying Zeppelins. They would be friendly rivals and planned to test their different methods against each other. Usborne had been looking into the Royal Air Force's response to aerial attacks by the German Air Force and he was all too aware of the amount of time it took for an aeroplane to start its engines, taxi, take off and gain altitude before seeking out enemy aircraft. Having pondered over this, Usborne devised a scheme to save those precious lost minutes. His plan involved mooring light aircraft to balloons anchored at certain heights. The mooring rope would be released when the aircraft was required for action. This, thought Usborne, would avoid the lengthy time required to get the aircraft to the necessary altitude. He was under no illusion as to the risk attached to testing such a plan and because of this he was adamant that only he and Commander Ireland, his chief volunteer pilot, would do the first test run. In the beginning everything went well, with the aircraft and its two pilots safely secured to the balloon and everything in position. At Commander Usborne's direction the mooring rope was released and the engine was started, but catastrophe struck when the tail of the aeroplane caught on the mooring-rope with the result that the aeroplane crashed to the ground killing both pilots. The accident had a devastating impact on Matthews who had grown to like and respect the Commander. As a direct result of the fatal accident the scheme was shelved and the work on the aerial torpedo was put on hold as no successor to Usborne could be appointed to the research facility.

Shortly after the fatal accident, further news about *Dawn* reached Matthews. Progress had been more rapid than expected and Lord Fisher had kept his word by arranging two further trials. The first trial was to test the selenium that was crucial to the working of the control pilot. The second was to test the vessel itself, where the boat would have to start, stop, turn to starboard, turn to port and fire a gun, all under the influence of just a searchlight. Lord Fisher had arranged for a test, on Penn Pond, in Richmond Park, London in front of leading government officials and scientists. A large shed was temporarily erected near the pond to house supplies and provide quarters for the night staff, but the pond was not specially protected or railed off: as the trial was taking place in late November and early December 1915, 'the park was almost deserted'.[14]

An agreement was drawn up by the Board of Invention & Research and signed by the Treasury. The terms were such as to offer little, if any, financial risk to the Chancellor of the Exchequer, Reginald McKenna, but in spite of this Matthews, full of confidence, was keen to press on for he saw no reason why a fair test could possibly fail. The agreement, signed on October 20th 1915, offered the following terms: that the test was to be carried out in a neutral laboratory and should J.J. Thomson, acting as the government's representative, be satisfied with the test then Matthews was to be awarded £25,000. Matthews and his staff would also then be considered as employees of the government with the inventor to receive an annual salary of £2,000. Regarding *Dawn*, Matthews, on each practical application of his invention, was to be paid a further sum until a maximum of £275,000 was reached. The Chancellor of the Exchequer was to be the sole arbiter of any payment.[15] The Admiralty offered to meet all the initial costs of instruments and equipment required for the demonstration which amounted to £500, as well as Matthews' personal expenses of £10 per week.[16] Lord Fisher requested that the terms offered to Matthews should be approved and 'not promulgated to any person whatever'.[17]

Professor Sir J.J. Thomson, famous for discovering the existence of electrons, the negatively charged particles of atoms, whilst working at Cambridge University in 1897, wrote to Matthews telling him that Professor Bragg had offered the use of his laboratory at University College, London, to carry out the testing of the 'pilot' and the all-important selenium. In the letter Professor Thomson had stipulated that once the experiments had started no adjustments could be made to the equipment. Although the conditions of the trial were strict, Matthews had complete confidence in his invention – after all it had performed well in the Edgbaston trial and he thought it would have no difficulty in doing everything asked of it. For the first trial, that of testing the selenium, Matthews got the selenium he wanted from Holland, as he felt it was of the best quality. During the test it had to withstand four different adverse environmental conditions: vibration, heat, cold and finally intense light. So to see how it fared, the selenium was placed next to a vibrating motor, put in a hot oven, frozen and then put under a powerful arc lamp. The selenium stood up well to the laboratory tests with its photoconductivity unaffected.

The second test took place in Richmond Park. E.E. Fournier D'Albe wrote an account of what happened in his book *The Moon Element*:

> The test was arranged for December 4th, but the pond froze over that night, and it had to be postponed to the 7th. When the great day arrived there was a bright sun all day long, and we all felt very nervous because it might affect the selenium and might do its own steering, although the pilot had been carefully shaded by means of circular discs of blackened brass spread over it horizontally. But the experts did not arrive until 4 o'clock, by which time the sun had sunk low enough below the trees to be well out of the way.
>
> We saw the cars coming over the distant ridges. I did the final tuning of the selenium cells on the boat and then went to look after the mine, half a mile away from the pond. The examining body arrived on the spot – a great array of talent, led by Lord Fisher. The Chancellor of the Exchequer and the First Lord of the Admiralty came, attended by a brilliant staff of leaders of British naval and military opinion. Mr. Balfour stood at the search light, and commanded the evolutions to be performed by *Dawn*.
>
> The little boat glided out from the small slip where she had been moored. She crossed the pond. A sweep of the searchlight, and she lighted her green light, turned to the left, performing a neat circle. Another touch of the luminous wand and she made straight for the shore, but was stopped in time by a warning glance from the governing beam of light. She then started again, and described figures of eight. For three-quarter of an hour the little boat careered twinkling about the pond, her red and green lights shining out alternatively. It was the prettiest play of fairy lights ever seen. She finally fired her gun and returned to her

moorings with only her selenium pilot at the helm. The test had been brilliantly successful, and the next day Mr. Matthews got his deposit of £25,000. He eventually worked the boat at sea at a distance of 3,000 yards in diffuse daylight, and up to five miles at night.[18]

There was some confusion as to which department should pay the £25,000, the Admiralty or the Exchequer? Matthews had written to the Chancellor on November 22nd 1915 asking him to clarify the situation and to express his desire 'to apply my invention to the practical benefit of my country with the least possible delay, and valuable time is being lost, owing to my not receiving instructions and the necessary facilities to proceed'.[19] Creditors were also pressing Matthews to settle outstanding bills which amounted to £15,000, including a payment to Dr D'Albe for all his assistance and guidance. Eventually McKenna took the decision that the Admiralty should make the payment. Throughout the trial of *Dawn*, Matthews underwent a series of gruelling interviews at the Board of Invention. His knowledge of selenium and electronics was tested to the extreme and the Board wanted to know his opinion about using the technology against Zeppelins. His collaboration with Fournier D'Albe meant that Matthews was a leading authority in the preparation, use and testing of selenium and his experience with radio-controlled airships gave credibility to his opinions. Matthews demonstrated another application of the technology when he demonstrated his mine exploder to a gathering of officials in that cold December. As with *Dawn* this had been a joint venture between himself and Fournier D'Albe. The collaboration between Matthews and D'Albe over the selenium mine lead to some animosity, for D'Albe claimed that he, not Matthews, had invented the device. But their friendship was an enduring one and they would continue to collaborate for many years to come. 'The mine consisted of a selenium relay that had a telescopic lens four inches in diameter mounted upon it and directed towards the searchlight station half a mile away.'[20] In order for the mine to detonate the searchlight had to be focused onto a small diaphragm thereby preventing any other searchlight from detonating the mine. With a mine 'set' ready for the demonstration the following eyewitness account gives details of what happened: 'We saw the beam of the searchlight swing round to where we were told the mine was placed. It touched the spot and instantly we saw the flash and the column of smoke.'[21] The secrecy of these trials was well kept as details didn't come to light until *The Moon Element* was published nine years later in 1924.[22]

Under the orders of the Admiralty, Matthews went to Portsmouth for further trials. For these trials the Admiralty proposed the loan of a hydroplane, *Cockleshell*, but after trying this Matthews thought it was unsuitable and wanted a slower vessel that, he thought, was far more conducive to a sea trial of his light-sensitive control gear. During one trial Matthews was, with a searchlight, able to control and manoeuvre a vessel at a distance of one and half miles at night. It was difficult to control the boat during daylight hours because special lenses were required for the receiver and Matthews was having some considerable difficulty getting the lenses just right. He wasn't completely happy with the sensitivity of the receiver either, for he realised that the searchlight beam had to be precisely aimed on the receiver. The lighter the conditions the less sensitive the receiver was to the searchlight beam. The proposed application to use the control unit for the remote detonation of mines was also scrutinised by the Admiralty. Again Matthews struggled to get the receiver to be sensitive enough during daylight but was able to detonate a mine at a distance of five miles at dusk. During night-time trials a successful remote detonation of six miles was achieved. Not happy with the daytime range Matthews altered the design of the lenses on the receiving apparatus and was, eventually, able to achieve successful results at a distance of nearly three and half miles during full daylight.

Technicians from the Admiralty witnessing these trials that took place on April 5th 1916 submitted a report to the Admiralty suggesting that the invention could be successfully used as a remote device to explode mines in abandoned trenches, thus preventing abandoned equipment from falling into enemy hands, and as a method of boat control over the ranges they had witnessed.[23] But the government also considered using the device to release asphyxiating gas in abandoned British trenches by day or by night (thus targeting enemy forces who had occupied them), direct a surface-propelled explosive vessel, or torpedo, by day, operate a small dirigible against airships and finally to detach bombs or large star shells from small balloons flown over enemy positions.

Trials had demonstrated that the device could be successfully used to explode mines and release asphyxiating gases in abandoned trenches. However using the device to control a torpedo was limited to the range of vision at which it could be manoeuvred. The torpedo would also have to remain on the surface on account of the necessity for the pilot or 'receiver' to be visible to the searchlights. For the same reason, its use to explode naval mines was not considered to be practical. Apparatus for more trials, planned to take place at Fort Blockhouse, was being prepared.

To demonstrate the versatility of his selenium device Matthews built several 'mine firing' sets 'with which small charges of T.N.T. were successfully exploded at Crayford in Kent on January 17th 1916 at a distance of some 1,100 yards in the presence of War Office representatives, who expressed considerable approval'.[24] However Lt. Col. Dumaresq 'was unable to recommend it for use under service condition'.[25]

King George V, already aware of Matthews' work with his Aerophone, made a visit to Fort Blockhouse to see the trials of the mine exploder. Vindicated by the trials, Matthews' work in the technology of remotely controlling sea-going craft and the use of long-distance explosion of mines was officially taken up by the government, but whether either was used in actual warfare Matthews was never to find out. His work on light-controlled boats was neglected and, instead of supervising the development of this technology, Matthews, along with his staff, 'was found other work to do'.[26] Matthews and Lord Fisher still remained close and would continue to collaborate.

The *Daily Chronicle* reported that Matthews found the trial of *Dawn* 'a thrilling and anxious time' with 'some of the Admiralty experts openly skeptical'.[27] The award of £25,000 was timely because at this stage Matthews' finances were in a poor state. With The Grindell Matthews Wireless Telephone Company having been declared bankrupt the previous year, he was in arrears to the tune of £15,000 and although he was awarded the

£25,000, in the end, Matthews never got the remaining £275,000 offered to him for his invention. The outstanding amount was never paid because according to various reports other means of fighting Zeppelins, such as the use of tracer bullets, had been adopted by the government. Matthews told the *Daily Chronicle* 'I started as a comparatively poor lad and that £25,000, with a good many more thousands of pounds, has been spent on developments in other directions.'[28] He was also later told by Lord Fisher that the amount of his salary, £2,000 pa, had come to the notice of various professors, who worked for the Board of Invention & Research and whose salaries were well known to be rather meager. Feeling rather put out and hard done by, they protested, and as a result Lord Fisher asked Matthews if he would withdraw his claim for a salary, which he did. In light of the fact that *Dawn* was never used by the British Government there have been claims that Lord Fisher used undue influence to ensure that Matthews was awarded the £25,000. There is no evidence of this and it would be an unusual act of a man with such a ruthless reputation of not suffering fools gladly.

Another invention that was the result of the close collaboration between Matthews and D'Albe was the 'Photophone'. This device allowed Morse signals 'to be telephoned by means of a buzzer applied to the ordinary service searchlight'.[29] With the war in Europe escalating, Britain was facing a new menace, the U-boat, from her enemies, and Lord Fisher asked Matthews to call in on him.

Notes

1. Barwell, p.50.
2. D'Albe, E.E. Fournier, *The Moon Element* (D. Appleton & Co., 1924), p. 69.
3. Barwell, p. 50.
4. Ibid., p. 51.
5. Ibid., p. 66.
6. Ibid., p. 69.
7. Ibid., p. 51.
8. Willmott, *World War I*.

9. Ibid.
10. D'Albe, p. 68.
11. National Archives T/1/11857.
12. Ibid.
13. Barwell, p. 53.
14. D'Albe, p. 69.
15. Barwell, p. 54.
16. National Archives T 1/11857.
17. Ibid.
18. www.openlibrary.org/details/moonelement002067mbp (accessed 12/07/07).
19. National Archives T/1/11857.
20. D'Albe, p. 71.
21. Ibid., p. 72.
22. Barwell, p. 55.
23. Ibid., pp. 56-8.
24. National Archives ADM 116/4766.
25. Ibid.
26. Barwell, p. 58.
27. Daily Chronicle, April 2nd 1924.
28. Ibid.
29. National Archives ADM 116/4766.

'The Grove', Winterbourne,
Harry Grindell Matthews' childhood home. (The author)

The Victorian Gothic splendour of the Merchant Venturers'
College, Bristol.
(http://www.10unitystreet.com/history.html)

Mrs Webb's School in the background just above the back of the horse to the left.
(Rosemary King – The Alverston Historical Society).

The Kursaal, Bexhill-On-Sea,
where Matthews succeeded in establishing a radio station.
(Society of Bexhill Museums)

LONDON HIPPODROME
Chairman ... ... Sir EDWARD MOSS
Managing Director ... ... OSWALD STOLL
Assistant Director and Chief of Staff FRANK ALLEN

No. 16.

Under the above number, in addition to the Items on the Programme,

THE WORLD WONDER!

A Wireless Controlled AIRSHIP

(The Only One in Existence)
WILL BE PRESENTED AND
Flown in the Theatre
BY THE INVENTOR
Mr. RAYMOND PHILLIPS

An Invention that must make the British Nation supreme in warfare.

Advertisement announcing Mr Phillips' demonstration of his airship.
(www.arthurlloyd.co.uk)

The 8th Earl De La Warr.
(Society of Bexhill Museums)

JACOB SAVERY,
AGRICULTURAL STORES,
RUDGEWAY.

Implements & Machinery,
Artificial Manures,
Agricultural Seeds,
Seed Corn, Feeding Corn,
Oil Cake, Meal, &c., &c.

Careful personal attention given to all orders by the Proprietor
JACOB SAVERY.

Samples, Prices, and other particulars on application.

Advert for Jacob Savery's Agricultural Stores.
(Rosemary King – The Alverston Historical Society)

British Electrical Exhibition held at the Kursaal in 1905.
Matthews can be seen seated on the left, fifth from the front.
(Society of Bexhill Museums)

HGM Portrait.
(ITN Source)

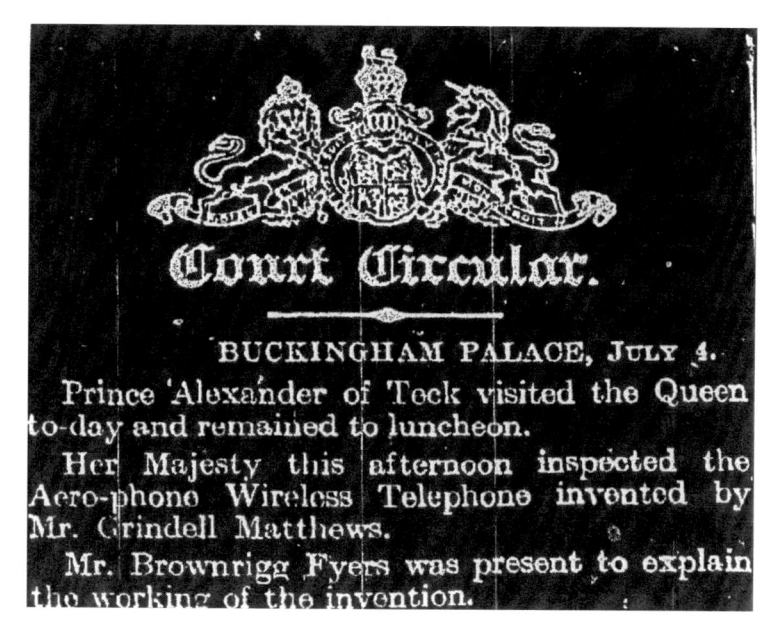

> **Court Circular.**
>
> BUCKINGHAM PALACE, JULY 4.
>
> Prince Alexander of Teck visited the Queen to-day and remained to luncheon.
>
> Her Majesty this afternoon inspected the Aero-phone Wireless Telephone invented by Mr. Grindell Matthews.
>
> Mr. Brownrigg Fyers was present to explain the working of the invention.

The Times Court Circular giving details of Matthews' demonstration of his Aerophone to royalty, July 5th 1912.

Raymond Phillips with his wireless controlled airship.
(Swansea County Archive)

The Western Mail receiving the world's first ever wireless press message in Cardiff. Matthews is standing on the left wearing earphones.
(Barwell, 1943)

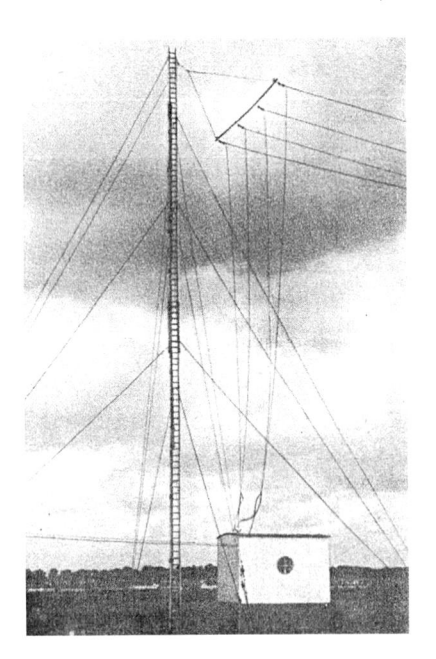

The radio station at Letchworth, 1913.
(Barwell, 1943)

The transmitting and receiving apparatus installed at Letchworth.
(Barwell, 1943)

Wireless communication with an aircraft in flight. Matthews can be seen on the left, seated at the transmitter whilst B.C. Hucks is at the controls of the aircraft wearing earphones.
(Barwell, 1943)

The New Passage Hotel on the banks of the River Severn. The workshops Matthews used can be seen on the right in the background. On the extreme right a transmitting aerial can also be seen. The long room shown on the left is where Matthews hung his airship from the ceiling.
(Eric Garrett)

N° 25,639   A.D. 1909

Date of Application, No. 25,639, 6th Nov., 1909
  ,,   ,,   No. 13,299, 1st June, 1910
Complete Specification Left, 6th June, 1910
(Section 16 of the Patents and Designs Act, 1907)
Complete Specification Accepted, 3rd Nov., 1910

PROVISIONAL SPECIFICATION.

No. 25,639 A.D. 1909.

Improved Means for Effecting Telephonic Communication Without Connecting Wires.

I, HARRY GRINDELL MATTHEWS, of Frieswood-Rudgeway, Thornbury, R.S.O., Gloucester, Electrical Engineer, do hereby declare the nature of this invention to be as follows:—

This invention relates to improved means for effecting speech communication by means of transmitting and receiving instruments without the aid of connecting wires, the installation being of a portable character.

The means for effecting the emanation of the waves which are required to travel undirected through space, comprise a hollow frame which may be of any convenient shape, circular, square, oblong or other closed figure. This frame is adapted to carry, distributed around its periphery, a plurality of bundles of iron wires, the lengths of which are disposed in the direction perpendicular to the plane of the frame, the several bundles being electrically insulated from each other.

Around the frame is also provided a continuous coil of many convolutions of insulated wire which may be within or may be wound around the beforementioned bundles of iron wire.

When it is desired to send a message, arrangements are made whereby can be switched into circuit with the said coil a battery, a telephone transmitter and a special device, to be presently described, whereby the battery current is rendered intermittent to an extremely high degree of frequency and alternatively, the battery, transmitter and intermitter can be switched out of the circuit, and a telephone receiver and relay switched into the circuit when it is desired to receive a message.

Preparatory to receiving the message, the circuit is made to include an aural, visual or other attention directing device arranged to be invoked by a relay battery which may be a portion or the whole of the battery by which the circuit is energised when a message is being sent.

The special intermitter of excessively high frequency comprises a pair of surfaces which are caused to travel relatively to one another at a comparatively low velocity, the electric current being transmitted from one surface to the other. The character of the surfaces is such as to normally interpose a resistance to the transmission of the electric current, and it is also such that the frictional contact of the rubbing surfaces causes an extremely momentary breakdown of the resistance, the resulting passage of an electric impulse re-constituting the resistance. This effect occurring at a multiplicity of points causes the electric flow to be intermittent with such a high degree of frequency that the voice waves produced in the telephonic transmitter are not modified in character but conveyed to the frame coil above described and originate space traversing waves possessing the same voice character.

[Price 8d.]

The dawn of wireless communication. Matthews' first ever patent for the Aerophone.

YOUNG INVENTOR WINS £25,000 IN ONE NIGHT.

Secret Experiment on Richmond Park Pond in War Time.

ZEPPELIN DESTROYER.

How a young inventor won £25,000 from the British Government in a single night as the result of an experiment with a ten-foot motor boat on Penn Pond, Richmond Park, is revealed to-day.

Hitherto it has been one of the closely guarded secrets of the war. The experiment, having for its ultimate object proof that an aerial destroyer, controlled from the ground, could be used to attack Zeppelins, was carried out with a model launch, in the presence of Lord Balfour, the late Lord Fisher, and a staff of experts.

It was so successful that a cheque for £25,000 was handed to the inventor, Mr. Grindell-Matthews, the next morning.

The Daily Chronicle. London. Wednesday April 2nd 1924.

The *Daily Chronicle* reports how Matthews got his £25,000.

Dawn.
(ITN Source)

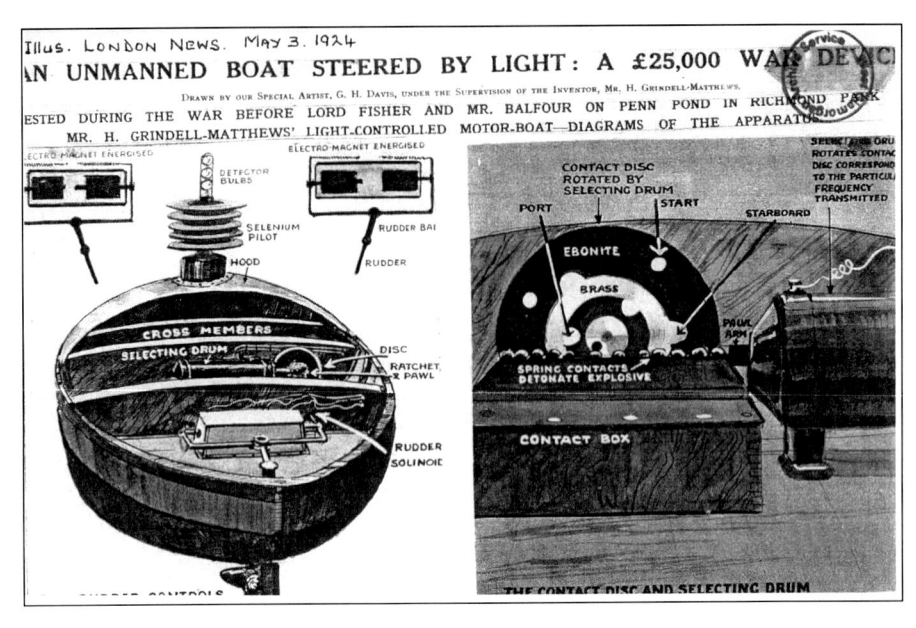

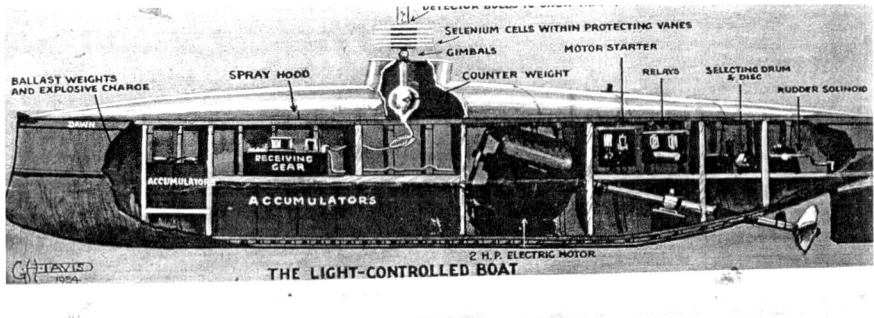

'The Unmanned Boat.' The *Illustrated London News*, May 3rd 1924, shows how *Dawn* worked.

CHAPTER 4

Into the Dark – Submarine Detection

During the First World War one of the biggest problems faced by the Allies was the savage threat posed by German submarines, or U-boats, to British merchant shipping making the dangerous journey across the Atlantic laden with vital wartime supplies. Extremely good at sinking Allied shipping, they were a deadly menace lurking beneath the cold, murky waters of the Atlantic, and they struck fear into the heart of many a British naval commander. The Germans were retaliating against the attempts of the British Navy at blockading supplies reaching Germany and embarked on a fierce U-boat campaign against the British Navy. Their tactical strategies were initially unsuccessful but in February 1915 Germany declared that her U-boats would sink ships without warning. In August that same year 185,000 tons of Allied shipping was sunk, with submarines being responsible for 149,000 tons. Germany was convinced that an effective U-boat campaign against the British would win them the war. Losses continued to mount with a total of 748,000 tons of shipping lost to U-boat attack in the second year of the war.[1] With the British press detailing the increasing losses and casualties caused by the underwater menace, Whitehall and the British public were starting to panic. Germany had no intention of letting up and British intelligence discovered that Germany had ordered the manufacture of a further 100 new U-boats.[2] Britain had to come up with an effective answer to the U-boat menace if she was to stay in the war.

Ultimately the Allies' answer to the threat of the U-boat was the convoy system where the merchant ships, with their precious cargo, would travel in large groups with an escort of destroyers. This, combined with effective air support, was a successful answer to the German menace. But in addition to such military tactics the backroom boys were busy developing a technological solution. The first technological leap in this field was a device that could detect the sound of a U-boat's engine under water, the Hydrophone. This was solely a listening device, similar to a microphone, and picked up sound energy under water and converted it into electrical energy. This had the major drawback in not being able to distinguish between friendly and enemy submarines. The term 'hydrophone' was first used by Paul Langevin and Constantin Chilowsky who pioneered the technology during the early twentieth century. The first recorded sinking of a submarine that was detected by a hydrophone was U-Boat UC-3 on April 23rd 1916 whilst on patrol in the Atlantic sea.[3]

The Canadian Reginald Fessenden invented the world's first sonar system in the United States in 1914. His device emitted a low-frequency noise that would echo off an object it struck and be detected by a separate receiver. Towards the end of the First World War ASDIC, also referred to as SONAR, was introduced in Britain. Developed by a top secret research department, headed by Mr R. Boyle, the **A**nti-**S**ubmarine **D**etection **I**nvestigation **C**ommittee, ASDIC was able to achieve, by the end of 1918, a detection range of 1,400 yards along with a reasonably accurate bearing.[4] The device was elegant in its simplicity, consisting of a combined receiver and a transmitter. The transmitter would send out a sound wave that would travel through the water. If the sound wave struck an underwater object the wave would be reflected and detected by the receiver and heard as a 'ping'. The duration of time for the reflected sound wave to be received was used to calculate the range of the submarine. The transmitter could be rotated in a similar manner to that of a searchlight, allowing the direction of the sound wave to be altered and the bearing of the object determined.

Matthews was well known to Lord Fisher, through his work on *Dawn*, and it was he who brought the problem of detecting U-boats to Matthews' attention: 'I am afraid that even the power of our Navy will be neutralized if the Germans continue their sinkings at such a rate.'[5] Well aware of Matthews' capabilities as an electrical engineer Lord Fisher suggested the Matthews think of a practical solution to the problem.

Submarines are driven by diesel engines which drive both the propellers and the generators that charge the batteries required to drive the electric motors. To be able to run its diesel engines the submarine must be on the surface or have its snorkel above the surface of the water. It is the batteries that power the electric motors allowing the propellers to be driven whilst the submarine is fully submerged, so when a submarine is fully submerged it is running on battery power. When the electrical motors are running, they create a magnetic field and therefore a magnetic disturbance. With this in mind Matthews thought that the magnetic disturbance could be detected from a distance. He wrote to Colonel Adams, on January 5th 1917, at the Ministry of Inventions, saying that 'I have recently devised a means of locating enemy submarines in a simple, rapid, and effective manner.'[6] But the Admiralty had previously carried out experiments along much the same lines and they found that 'no (magnetic) field of any practical value extended beyond a few feet from the hull of a submarine'.[7] The Admiralty had been trying to solve the problem of submarine detection for some time and the assistant Comptroller at the Ministry of Inventions, W.H.D. Clark Major, wrote to Matthews later that same month:

> Experiments carried out by the Admiralty to determine whether the screening effect of the plating of a submarine was great enough to prevent the transmission of magnetic disturbance ON WHICH YOUR METHOD DEPENDS FOR ITS EFFECTIVNESS. The screening effect is so great as to render the method impractical, and as a result the Admiralty regard it as useless to proceed further with the matter. The Admiralty see no further step that they could usefully take as regards your proposal.[8]

There had been a delay in Matthews receiving this letter: a delay which the Admiralty explained was the result of them consulting an expert regarding his proposal. But Matthews discovered that the Admiralty had duplicated his experiments without giving him the opportunity of being there to explain the workings of his own device and that, during the experiments, they had used a 'Brown Relay'. He 'knew it was quite impossible to work his apparatus with that type of relay and, regarding the screening effect said that he had effectively satisfied himself how this important point can be overcome'.[9] Matthews suggested another trial so that he could demonstrate his apparatus and prove that it 'was indeed practical' and 'how the difficulties experienced by the Admiralty authorities can be surmounted'.[10]

The inventor was keen to show how good his device could be when operated correctly but with breathtaking shortsightedness Colonel Adams wrote to Matthews:

> The view of the Admiralty is that experiments have confirmed that there is practically no external fluctuating magnetic field produced by the motors of a submarine. As your proposed means of detection depends on such a field for its operation, it is regarded as inapplicable to this particular problem.
>
> Brown Relays may not have the amplifying powers of a series of electrical valves, it is thought that the amplification must have rendered audible any disturbances that could conceivably be detected under the far more disadvantageous conditions under which your apparatus would be required to work. You will not have failed to note that the cable loop in these experiments was close up to the hull of the submarine. It would not be unreasonable to suppose that if the disturbances were great enough to effect your purpose they would have been detected under such favourable conditions without any amplification at all.
>
> In view of these results the Admiralty can see no further step they can take with regard to your proposal.[11]

However, to his credit, Colonel Adams did suggest that Matthews place the matter 'before the Board of Invention and Research and if they see any possibility of success they would be in a position to arrange a trial'.[12]

Convinced that he had solved the technical problems the Admiralty had experienced with submarine detection, Matthews found all this incredibly frustrating and on February 6th wrote back to the Admiralty saying that he acknowledged the Admiralty's position and that 'experiments done by the Admiralty confirm results already arrived at, namely that there is practically no external fluctuating magnetic field produced by the motors of a submarine'. But he went on to say that he was satisfied that there was an external fluctuating magnetic field produced and offered to give a demonstration of his device, concluding 'it is difficult for me to understand why I have not been given the opportunity'.[13]

Matthews' dogged perseverance paid off when a week later he received another letter from the Board of Invention & Research stating that his 'proposed method of detecting submarines' had 'been forwarded by the Comptroller of Munitions and Research' and to say that they 'would be glad if you could find it convenient to call here and give an outline of your proposals'.[14]

At last! A chance to demonstrate his submarine detection apparatus. With the help of Lord Fisher, later that month Matthews got the opportunity for the trial he had so desperately wanted. On February 28th 1917 Matthews, along with his assistants Mr Wheeler and Mr Lynes, left London and made the long journey by train and motor car to Portsmouth Harbour. After having seen all the equipment safely stored in the station cloakroom, they retired to the nearby Keppel's Head Hotel, overlooking the Harbour.

An unnamed author of a document held in the Swansea Archives[15] gives details, and offers a glimpse, of that very first sea trial undertaken by Matthews and his team in March 1917: on March 1st, an optimistic Matthews met with the Flag Captain of Portsmouth Harbour and Commander Crowther from the Admiralty, to explain exactly how he planned to test his apparatus. The Drifter HMD *Faithful* was made available for the trial and this was placed at his disposal for as long as he required. Sandown Bay on the south-eastern side of the Isle of Wight was chosen as the best place to carry out the trial. The Admiralty had provided a submarine which was moored in the Bay in readiness for the trial to begin and which 'would be at their disposal for Friday, Saturday, Sunday and Monday'. Matthews was delighted with all the arrangements and later that day, accompanied by Lynes and Wheeler, went to collect their equipment from the railway station.

On the following day they started the arduous task of laying the copper cables in the bay. When a DC current is passed along a wire, or a cable, a magnetic field is induced around it, a phenomenon known to all school children when they see the effect the current flowing through a wire has on a nearby compass. Any disturbance made to this magnetic field would indicate the presence of an electric motor and therefore a submerged submarine. A similar effect can be seen when a hairdryer is switched on near a television. In the trial, the disturbance to the magnetic field was to be relayed to the electronic detecting apparatus installed on board HMD *Faithful*. Progress was good until, having nearly completed the cable laying, the skipper was ordered to shut down the engines by the captain of a motor launch which was accompanying another submarine on hydrophone tests. Annoyed by such as request they stopped work for a while but Matthews, fully aware of how precious time was, shortly began laying the second cable. But the motor launch quickly reappeared with repeated threats from the captain that he would report the skipper for running her engines without his permission. Despite this wrangling the crew managed, after several hours, to finish laying both cables across the bay.

With all the detecting apparatus installed on board HMD *Faithful*, and the cables in place, Matthews requested that the submarine be brought into the bay and it 'made several passes and dives, over the cables, without giving any indications as to her intentions'.[16] Matthews, onboard HMD *Faithful*, was crouched over the instrument panel listening intently for signals declaring the presence of the submarine's electric motors. After making several adjustments to the equipment, suddenly through his earphones, Matthews heard the tell-tale sound, and excitedly confirmed that he 'heard the starting and stopping of the submarine's (electric) motors'.[17] The following day brought rain and bad weather and so the day's trial had to be abandoned. But Lynes had managed to hear 'sparking' from engines of the submarine when it was on the surface. Commander Crowther recorded that 'important results had been obtained'.[18] Then the weather started to take a turn for the worse: so much so that HMD *Faithful* did not leave the harbour, leaving Matthews with noting else to do but check all the equipment.

Matthews left the Keppel's Head the following morning to find the weather much improved but was told the submarine was not available for trial. So he decided to make some local tests. These showed that his equipment was sensitive enough to detect the sparking from the engines of passing motor boats. Such was the sensitivity of the equipment that the stopping and starting of electric trams running along the seafront in Southsea at a minimum distance of 200 to 250 yards was also heard. The electrical activity of the trams' motors were disturbing the magnetic field generated by his apparatus and it was this disturbance that Matthews could hear. He was thrilled; his equipment was showing real potential.

When he returned to his lodgings later that day, the hotel receptionist handed Matthews an official-looking envelope and he realised immediately that it was from the Admiralty. They wanted to know when he 'would be ready to demonstrate the results of his experiments' and informed him that HMD *Faithful* was 'shortly required for patrol duties'.[19] Anxious to hang on to the destroyer, he immediately wired a reply giving details of the results they had got and stating that he wanted to continue trialing his apparatus. But to his utter frustration the spell of bad weather continued for the next couple of days, making the use of the submarine impossible. But despite this they again repeated the local test, only this time they got clearer results. Progress, however small, was very welcome. During this repeat experiment special attention was paid to the type of signals they detected. Matthews managed to persuade the skipper of HMD *Faithful* to sail out into the harbour so that he could just extend the range of detection. Sailing past Fort Blockhouse Matthews discovered that the audible note received from the electrical generators on Fort Blockhouse and was at its loudest when they were directly opposite them. His apparatus was able to detect the running of the generators from a distance of three-quarters of a mile. Throughout the tests, such was the sensitivity of his equipment that the individual pulsations from the electric motors of nearby trams were also heard.

Later that evening Matthews' assistant, Mr Wheeler, made a telephone call from their lodgings to Captain Walwyn, at the Admiralty, and told him of the encouraging results they had achieved, despite the bad weather, over the past few days. He mentioned that Matthews would be ready very shortly to give a demonstration of his working apparatus to the Admiralty. They all kept their fingers crossed in the hope that they would be allowed to hold on to the submarine for a bit longer.

With depressing predictability the weather on the 8th was too rough to do any further trials and Lynes and Mr Child, one of Matthews' technical assistants, did the routine testing and checking of all the instruments. Later that afternoon Commander Crowther told Matthews that the submarine was required on the following day by the Admiralty and wouldn't be available for another trial. Matthews' anxiety only grew with the weather being too bad for any testing to be done for the next two days. With time slipping away, all he and his team could do was stay on board HMD *Faithful* testing the valves, amplifiers and equipment. The work was frustrating and arduous. Every evening the team would have their supper at the Keppel's Head and discuss the results, how improvements could be made to the apparatus, and keep an anxious eye on the local regular weather reports.

By March 11th the weather had improved with 'visibility half-a-mile' and arrangements were made to resume testing. However, after some difficulty in getting HMD *Faithful* out of the harbour, owing to the fog, the skipper told Matthews that had received instructions to return immediately to port. Matthews couldn't believe it but it was all about to get much worse for during the return trip back to port the fog closed in and *Faithful* had a minor collision with another vessel, causing slight damage. But eventually the skipper got her back and safely moored in port just before lunchtime. Matthews decided at this point to retrieve the cables from the sea bed and allow them to dry out, thereby restoring their sensitivity and giving him a chance to inspect them for any damage. So with no submarine available, damage to HMD *Faithful*, and bad weather Matthews wasn't having much luck but had achieved some encouraging results. On the 14th Matthews, along with Wheeler, visited the Flag Captain to give him a document containing all the results from the trials. By this time Matthews had had the use of HMD *Faithful* for a fortnight and the Flag Captain did not have the authority to continue this arrangement. Between them they decided that Wheeler should visit the Admiralty back in London with the document containing all the results obtained in the trials. Wheeler was also to try and get permission for more time to use HMD *Faithful* and request a postponement for Matthews' demonstration of his apparatus that had been arranged for Thursday 15th. Having met with Admiral Duff at the Admiralty, Wheeler wired the Keppel's Head stating that Duff couldn't authorise the continued use of HMD *Faithful*, but in light of their results, the Admiralty was prepared to make arrangements for further trials at Bristol and later Barry, in South Wales. So with things not having gone quite as planned, but having got some good results with their apparatus, an exhausted Matthews packed up his equipment and returned to London to prepare for the next trial.

The second trial at New Passage in Bristol happened some time between March and May that same year, 1917. What happened during this second trial, is not recorded but details of the third trial, at Barry in June, were recorded by Matthews' chief research assistant B.J. Lynes.[20] Lord Fisher had arranged the use of HMD *Sea Queen*. But Matthews learned that he was to work with a person mysteriously referred to as 'Naval commander A' in Barwell's biography. This commander was in fact Commander Crawford whose name appears in Lynes' report on the trials carried out at Barry. Crawford was, apparently working in the same field of research on behalf of the Admiralty. But the question as to why Crawford was to be there and what his exact motives were, was unclear. On May 30th all the necessary arrangements had been made for Matthews along with Lynes and Commander Crawford to travel from London to Barry on the south coast of Wales, a few miles from Cardiff.

Little was Matthews aware but this third trial of his detection apparatus was to turn into a farce and would serve only to deepen the feelings of suspicion and mistrust he had for the British Government. After having arrived in Barry, Matthews' team placed all of their equipment and instruments on board *Sea Queen* and made the necessary preparations to get everything ready. Whilst they were unpacking their equipment an officer by the name of Lieutenant Manson arrived with his own equipment and declared that he was there to work with Commander Crawford. June 2nd saw Matthews, Lynes, Crawford and Manson board *Sea Queen* to begin the trials. But Commander Crawford requested that they all go to Swansea for stores as he needed further equipment and as they returned so late in the evening no experimenting was done that day – and the following day, Sunday, Commander Crawford declared a 'rest day'.

By Monday 4th they at last managed to make a start by tuning up the instruments and with the use of a small electric motor, the signals from which they used as a standard comparison, Matthews gave a demonstration to both Manson and Crawford showing them exactly how their equipment worked. Manson was, with his own equipment, able to get equally good results.

Tuesday began to see some encouraging results with their first trial made in deep water. The submarine, on loan from the Admiralty, was launched out into the Bristol Channel and Manson, using his detection apparatus and working closely with Crawford, 'asserted that they heard the submarine starting and stopping her motors'. Lynes, with Matthews' equipment, 'was able to distinctly hear the submarine's motor'. When he looked out of the cabin of *Sea Queen* he saw that 'the submarine was totally submerged'. When it did surface, 'within the next two minutes, she was one and three quarters miles away'. Crawford also reported that he could hear the submarine's electric motors at the same distance and confirmed the statement made by Lynes 'that the motors of the submarine could be overheard at a distance which was estimated at eleven and a half miles, but later due to defective valves, results were not so satisfactory'.[21]

On Wednesday Matthews 'decided to make local experiments in the basin alongside the submarine'. They used their standard electric motor which 'gave off good signals' when on the deck. However, when they placed the motor into the submarine's conning tower 'all signals at once vanished'. They kept repeating the experiments but could not pick up any signals. This was the problem of the screening effect caused by the plating. Despite this unexpected result they moved the hull of the *Sea Queen* 'from being in metallic contact to the submarine' and stationed the ship at a distance of about 30 yards. But they still couldn't detect any signals from the electric motor placed inside the conning tower of the submarine. Using his own equipment Mr Manson could not detect any signals either. Matthews thought that this was due 'primarily to the weakness of the valves' that they were using.[22]

They spent the whole of the following day installing the electric motor into a sheet metal tank in which they could completely encase it to examine, more closely, the screening effect. Later that same night they tried some testing but all they managed to do was damage the instrument valves and batteries of the Audion amplifier. The Audion amplifier is a device that amplifies very weak electrical signals making them more audible. Lynes had to spend the whole of the next day 'overhauling the instrument and reconnecting the batteries' while Matthews went into the local town to get some new valves to replace those that had been damaged. He realised that the valves used by Lieutenant Manson had been manufactured in France and were of a far superior quality. However these were only available in limited numbers and the ones that Lieutenant Manson had he was naturally keen to hold on to. Lynes tested the valves which Matthews had managed to purchase, but found them all to be 'very noisy and inefficient'. Despite this they pressed on and did further tests with the motor encased in the sheet metal box. At first they made a test from close quarters with the top off but could only hear a slight signal 'about one-tenth the signals at New Passage', that is the previous trial at Bristol. However Manson reported that he 'was able to hear at greater distances'. Lynes recorded his disgust when he found out that Commander Crawford had invited his wife on board, showed her what was happening, and asked her to listen to the signals with Manson's equipment.

The submarine was unavailable on the 12th and for the whole of the following day Matthews' team was 'occupied by testing various ideas of Commander Crawford'. Consequently 'no work of real value was done as he monopolized the submarine and boat'. Manson left for London on the morning of June 14th whilst Matthews and his team pressed on with their own experiments. Had Manson gone back to the Admiralty with details of what he had witnessed? Crawford, now working without Manson, did another trial and whilst travelling out to the submarine in the motor launch he met the submarine returning to port and 'was able to hear the motors running'.[24]

Not until Monday 19th when the replacement valves Matthews had ordered arrived was he able to resume testing. He was well aware that the inferior quality of the valves they were using was hampering their results and was desperate to get the French type that Manson was using. 'Strenuous efforts were made to get some French valves and the help of the Admiralty was enlisted and although they promised a supply not a single valve ever materialized.'[25] Matthews thought that fitting the new valves to the Audion amplifier would help overcome the inferior sensitivity of the valves he was using. However, he wasn't able to fit them until later that afternoon of the 20th as, contrary to the agreed programme, Crawford was out on the *Sea Queen* all that morning. It was too late anyway as time had run out and the trial had to end. Matthews was disappointed to say the least.

After the trial at Barry was completed, Matthews wrote to Commander Yeats Brown at the Admiralty:

> I wrote to Captain Walwyn stating that I had returned from the Bristol Channel and that the modifications I had carried out to my apparatus had produced most important results. On May 13th you asked to see me and requested I return to New Passage, Pilning and continue my work. On the following Sunday Mr. Wheeler telephoned you stating that I was ready for your promised visit and you came down with him on the following morning. In the afternoon we went out to the 'Sea Queen' lying about one and half miles up river, when at varying distances up to 800 yards you heard from the 'Sea Queen' a two horse power motor placed on the 'Volunteer'. On Tuesday Morning further tests were made, as a result of which you expressed yourself satisfied and decided that the tests with a submarine should be made and suggested Weymouth off Lulworth Cove [*sic*]. On Friday, May 25th, Mr. Wheeler telephoned you, and you stated that it was almost certain that Barry would be the place selected for the tests. That same day I received a letter from you confirming the arrangements and that a submarine would be available for me on all days of the week except Tuesday and Fridays and you thought that both myself and Commander Crawford would have all we required to make a start. You also stated that you would get me a new valve set and, of course, anything else I wanted.
>
> As a result of a week at Barry I had come to the conclusion that the work was being carried out under most disadvantageous conditions.

In the first place, for tests of this description it is essential that they should be carried out without disturbing surroundings, such as exist at a busy and noisy centre as Barry. Further it is not feasible to attempt to experiment on more than one method of test at one time.

I therefore called on you on Thursday, June 14th, and explained to you the difficulties, and my views were supported by Lieutenant Manson, who had come up from London, but you stated that the arrangements could not be altered, and I returned to Barry, having ordered certain valves which, however, proved useless for the work.

I discussed the position fully with Commander Crawford with the result that he agreed with me that it was advisable I should take my apparatus on the 'Sea Queen' to New Passage, and work on the results I first obtained on the occasion of your visit.

It was necessary for me to obtain further valves similar to those Lieutenant Manson was using, and on Tuesday, July 10th, I telephoned you requesting the facilities for obtaining these, and you replied that Lieutenant Manson must obtain them, and telegraphed Commander Crawford for Lieutenant Manson to come up to London for this purpose.

On Wednesday, July 11th, Lieutenant Manson and myself called upon you, and you stated that as no tangible results had been obtained during the five weeks, the 'Sea Queen' would be sent back to Bristol, and further experimental work stopped and you could give me no facilities whatever for obtaining the valves and instruments I required.

Though the results at Barry have not been such as you may have anticipated, I have explained to you the difficulties that I have had to contend with.

The results, even where [sic] negative, have at least the advantage of showing what methods may be ruled out, and are, therefore, of value.

The results of my research, as witnessed by you at New Passage, surely justified every possible means being tried to prove this conclusively.

I know the patience and trouble required to achieve success.

The facilities I require are the use of the 'Sea Queen' at New Passage and the valves and instruments, and I shall be glad to know definitely whether or not I am to be supplied with these, when I will at once proceed with the work.

Yours faithfully,
H. GRINDELL MATTHEWS[26]

Matthews' frustration is only too apparent. The first trial in Bristol had given some really encouraging results and he obviously felt that the Barry trial just wasn't conducive to effective experimenting. His frustration at the Barry trials is evident in the letter he wrote and he felt that the Admiralty was, far from assisting him, hampering his work.

After writing this letter to the Admiralty Matthews travelled to Cherbourg, in France, 'where he had been promised full facilities by the French Government'[27] to continue his work with submarine detection.

On September 5th 1917 an inquiry, presided over by Mr Justice Clauson, was held into the events that took place during the trials at Barry and Matthews was summoned to attend. Also present at the inquiry were Mr W.L. Hichens, as chairman, Sir R. Peirse and Sir H. Skinner. Mr Hichens' inquiry was set up to investigate the allegations contained in the written reports of Commander Crawford and Lieutenant Manson and to report whether or not Mr Grindell Matthews had offered money to Commander Crawford and Lieutenant Manson in order to obtain possible pecuniary benefit from researches conducted by them whilst working for the Admiralty. The inquiry was also to consider whether or not Commander Yeats Brown had obstructed, at the suggestion of Matthews, Commander Crawford during the course of his research and whether Crawford claimed, as his own, the results achieved by Lieutenant Manson. Finally the board was also to consider the manner in which the whole Barry trial was organised and whether it was done so in a manner to offer the best results.[28]

Matthews appeared at the inquiry to answer all the questions put to him and 'left the inquiry confident that the matter would be resolved'.[29] However he was never told of the board's findings and suspected that he had 'a black mark against his name'.[30] Indeed a file held in National Archives states that 'a Committee of Enquiry held in 1917 found that he had been guilty of offences under the Prevention of Corrupt Practices Act'.[31] Years later in 1921 Matthews received a letter from Commander Crawford, who was, by then, retired from the Admiralty and working on underground wireless systems. In the letter, after congratulating him on his 'success with film records' he goes onto say 'I understand that the Admiralty treated you very badly over the Barry experiments and I want to know why. I have a suspicion that there was some dirty work somewhere; much of this could be cleared up if I could meet you I think.'[32] But there is no record of whether they met or not.

At the beginning of 1918, a letter from the Admiralty dropped onto Matthews' door mat at Bickenhall Mansions, Gloucester Place in London. Opening the letter he read that the Admiralty thought that 'no further benefit is likely to result either by continuing these experiments or by prolonging the correspondence on the subject'.[33]

So that was that. It was over as far as the British Admiralty was concerned. Government files show that the Admiralty had been forbidden to correspond, or have any further dealings, with Mr Grindell Matthews. Matthews still retained complete faith in his submarine detection apparatus and turned his attention to the French authorities who had offered him help and further assistance with his work. A fluent speaker of French, Matthews knew the editor of *Le Journal* and it was 'through the editor of *Le Journal* [he] was offered facilities for continuing with his experiments under the patronage of the French government'.[34] He funded these experiments out of his own pocket but 'materials for his experiments were supplied free of charge at Cherbourg'.[35] When he arrived at Cherbourg he was also introduced to Commandant Marguery, the officer in charge of the French Arsenal. Matthews was given every assistance whilst at Cherbourg with the yacht *Crotale*, donated to the French Government by the American actress Florence Turner, being made available for his use. The officer in charge of the patrol at Cherbourg was appointed to 'evaluate the results and findings of Matthews' investigations'.[36]

On February 15th 1918, the *Crotale*, packed with all his necessary equipment, sailed out into deep water. A submarine, on loan from the French authorities, was placed in position. The use of the submarine was limited because it was also being used for other submarine detection trials. With permission given by Marguery, the *Crotale* sailed to within a three miles radius of the submarine, who had her electric motors running at full. Matthews tuned his apparatus and set about detecting the magnetic disturbance caused by the submarine's engines. However the distance was too great, even though, in earlier Barry experiments engines had, according to Barwell, been heard at eleven and a half miles.[37] Later that day Matthews was able to detect signals from a distance of 1,000 yards. Following this, the *Crotale* sailed a circular path around the submarine, whose engines were started and stopped repeatedly, and Matthews was able to hear signals 'with excellent distinctness'.[38]

Flushed with this success, Matthews was given, for a period of three hours, the use of a captured German submarine, EB26. The experiment was, however, a failure. Matthews thought that the earth electrodes were the problem. The sea was so rough that with the 'electrodes jumping on the waves, all that could be heard was a great deal of noise'.[39] Matthews, ever the inventor, was able to design a new type of electrode that operated under the water. Despite bad weather hampering the experiments over the next few days he, along with his assistants, was able to 'hear the submarine as she slid out from the harbour into the channel'.[40] Unfortunately an accident happened when a propeller of one of the vessels used in the trial, was fouled by the cable used to generate the magnetic field.

Results from a trial on March 7th were so encouraging that Marguery 'promised the use of a torpedo boat'[41] in addition to the submarine, for more trials. Marguery wrote in his report full details of the trials, the results Matthews had achieved and how they were progressing. The report suggested that Matthews be given the use of a much calmer base, such as Toulon or Brest, so that the trials were not at the mercy of the weather quite so much. Later in April Matthews received notice that Commandant Cadon, from the British minesweepers, had arranged for the trawler *Daniel Harrington* to be used for further trials. This afforded him the opportunity of setting about how to eliminate noise from the propeller causing interference, which after much work, he managed to achieve. On the next day he was able to 'obtain signals from a French submarine, both on the surface and whilst submerged, up to a distance of 500 yards, free of background noises'.[42] In May after more trials, Matthews obtained 'perfect signals' only to be followed by poor results due to bad weather. The trials were arduous and physically demanding. Being at sea for long periods, often in rough weather, started to take its toll and Lynes fell ill and spent several weeks in hospital where he remained until the end of June. However, in the meantime, Matthews was able to increase the range of detection 'from his previous French record of about 1000 yards to a distance of which approximated nine miles'.[43] Working without Lynes the strain on Matthews' health increased and, inevitably, he fell ill as a result of the dreadful weather and poor working conditions on board *Crotale*. Laid up for a few days Matthews pondered the team's progress and he was pleased with what they had achieved: detection over a distance of nine miles. Progress continued at a steady, yet definite, pace and with new improved detecting apparatus designed to be more sensitive, reliable, and less susceptible to the weather, better results were just around the corner when suddenly Matthews received a summons to visit London.'

The summons was to see Lord Fisher and the outcome of what was to follow was an invention of Matthews' called the 'Bullet Circuit Closer'. Details of this in Barwell's biography are vague to say the least. The invention came about after a conversation with Lord Fisher about the problem of Allied tanks getting stuck in no man's land and consequently falling into enemy hands. The Bullet Circuit Closer was used to destroy the stranded tanks in the event of an Allied retreat. With no registered patent or any publication giving details of this invention it is difficult to determine how it worked. Both the British and the French authorities were interested in the device and a demonstration was given to the French at Montreuil on September 6th. After announcing they were satisfied with the Bullet Circuit Closer, Matthews received the following letter from the War Office:

General Headquarters,
British Armies,
France.
Nov. 1, 1918.

Dear Mr Grindell Matthews,
    With reference to a letter from Paris, which was not sent by despatch rider, and has only just arrived. The official application was sent to the War Office, saying that we could make use of your

Circuit Closer, and asking that arrangements might be made with you for the use of your invention.

Yours sincerely,
(Signed) J. E. Edmunds.[44]

Matthews was to learn on November 8th that the War Office had ordered 35,000 Bullet Circuit Closers with him receiving a royalty of five shillings for each one made. However just three days later the Armistice was signed and there was no longer a need for the Bullet Circuit Closer.

Relieved that the nightmare of the war was over, and weary from strain of the past few months, Matthews returned to Cherbourg to collect his equipment and take it back to Pilning. He was exhausted. The Allies were euphoric that 'the war to end all wars' was over and as church bells rang, a bewildered Britain was rejoicing. The British and French governments were quick to drop Matthews and his submarine detection apparatus. No longer interested in it, with the war behind them they wanted to concentrate on rebuilding their economies and prepare for peace, not get involved with inventions that had military applications. However Matthews was not alone in thinking that the Germans were not beaten and suspected that Europe would in the near future be engulfed in another world war. In the 1930s, with Germany rattling her sabre, Matthews would again pick up his submarine detection apparatus only to encounter the same official short-sightedness. In the meantime, back in his London flat and enjoying a rare moment of relaxation, Matthews pondered his future and through his association with Fournier D'Albe turned his attention to 'talkies' and the recording of sound onto film.

Notes:

1. Willmott, *World War I*.
2. Ibid.
3. www.ob-ultrasound.net/hydrophone.html (accessed 18/03/07).
4. Ibid.
5. Barwell, p. 62.
6. Swansea Archive D/DZ 346/13.

7. Ibid.
8. Ibid.
9. Ibid.
10. Ibid.
11. Ibid.
12. Ibid.
13. Ibid.
14. Ibid.
15. Ibid.
16. Ibid.
17. Ibid.
18. Ibid.
19. Ibid.
20. Ibid.
21. Barwell, p. 63.
22. Swansea Archive D/DZ 346/13.
23. Ibid.
24. Ibid.
25. Barwell, p. 63.
26. Letter from Matthews to Commander Yeats Brown, undated. Swansea Archive D/DZ 346/13.
27. Barwell, p. 64.
28. Swansea Archive D/DZ 346/13.
29. Barwell, p. 64.
30. Swansea Archive D/DZ 346/13.
31. National Archives ADM 116/4766.
32. Barwell, p. 63.
33. Ibid., p. 64.
34. Ibid., p. 65.
35. Ibid.
36. Ibid.
37. Ibid.
38. Ibid.
39. Ibid.
40. Ibid.
41. Ibid., p. 66.
42. Ibid.
43. Ibid., p. 67.
44. Ibid., pp. 69-70.

CHAPTER 5

'You Ain't Seen Nothin' Yet!'

BEFORE THE ADVENT of 'talkies', when a film was made the images and the sound were recorded separately, with the images being recorded onto film and the sound onto a disc or phonograph cylinder. When played back, the sound would never run in synchronisation with the pictures, resulting in a really strange effect. Matthews foresaw the growing popularity of cinemas and proposed to solve this problem by recording both the sound and images on to the same reel of film. This meant creating an optical sound track to run alongside the photographed images so that they would never run out of synchronisation. Progress had initially been made in Germany at the beginning of the twentieth century but still no practical means had been developed to record both sound and images onto the same reel of film successfully.

The French inventor Louis Aime Augustin is thought to have recorded the world's first moving images in October 1888. Lasting just a few seconds he recorded the busy horse-drawn traffic crossing a bridge in Leeds.[1] Augustin was to mysteriously disappear in 1890 when he boarded a train in Dijon, France on September 16th 1890 and was never seen again. A remarkable inventor and chemist, like so many others, he was never to receive the recognition he deserved.

It wasn't until a few years later that the public first saw the magic of moving images in Kinetoscope parlours. Invented by William Dickson, the Kinetoscope was a large, coin-operated contrivance made out of polished wood with a viewing piece on the top. Working for a time with Edison, Dickson invented the

Kinetoscope in the early 1890s. Driven by an electric motor, the device wound a continuous length of film in front of an electric lamp. The winding film, with a series of sequential images recorded upon it, would give the impression of a moving image when viewed through an eyepiece on top of the machine. The images were recorded onto the celluloid film with a device called the Kinetograph an early type camera with a gate and a shutter. Kinetoscope parlours were popular attractions towards the end of the nineteenth century and gave people their first experience of motion pictures. The drawback of the Kinetoscope was that the images could not be projected onto a screen, a problem the Lumière brothers solved in 1895. Called the Cinématographe their movie camera recorded and projected the images and was used to show the worlds' first public movie, *La Sortie des Ouvriers de L'Usine Lumière*, in Paris in 1895.[2]

Silent films were initially seen as quite a novelty but the stumbling technology advanced during those early pioneering days and the film industry was to burgeon into a multimillion-pound industry. With live accompaniment provided by a piano, audiences flocked to crowded cinemas, to watch the antics of movie stars like Charlie Chaplin and Harold Lloyd. Those early silent films with their scratchy, jerky movements were hugely popular, providing escapist entertainment for the cinema-going public.

Colour came much later to film and was added, in the beginning, by painstakingly hand painting each individual frame. Later black and white films were photographed through different coloured filters giving the illusion, albeit a relatively poor illusion, of colour. Technicolor, first used in the 1930s, employed a technique that involved adding coloured dyes to the film before photographing through different coloured filters. It is worth mentioning here the fascinating life of William Friese-Green.

Born in Bristol in 1855, Friese-Green was one of those wonderfully endearing, romantic, inventors of the nineteenth and twentieth centuries and like Matthews was never to get the credit or recognition he deserved for his pioneering work with moving images and photography. A romantic account of his life, *Close-up of an Inventor*, by Ray Allister[3] was turned into film, *The Magic Box*, for the 1951 Festival of Britain.[4] A passionate advocate of the British film industry, Friese-Green is considered to be the inventor of cinematography, the recording of moving images for cinema; he patented different cameras along with various methods of photographic printing. In 1889 he went on to patent the world's first motion picture camera. In his youth he was apprenticed to a photographer and later went on to set up his own photographic studios. After moving to Bath he met John Rudge. Known as the 'Wizard of the Magic Lantern' Rudge experimented with various types of magic lanterns and invented the 'Phantascope'[5] a device that projected seven slides in quick succession thereby giving the impression of motion. Rudge, working with Friese-Green, improved the instrument so that it could project colour images and they called their device the 'Biophantascope'.

By the mid 1880s Friese-Green was experimenting with photographing images onto celluloid film and later he invented the chronophotographic camera that used perforated celluloid film to record images. Unfortunately his chronophotographic camera never caught on due to its poor reliability. Undaunted he went onto experiment with stereoscopic or 3-D cameras but again success eluded him. Later, declared bankrupt, and after a brief spell in prison over a matter relating to his financial difficulties, he sold the rights to his chronophotographic camera to help alleviate his straitened circumstances. Never one to admit defeat he went on to experiment with colour images and invented a system called Biocolour. This was a rival to a system invented by George Smith and Charles Urban called Kinecolour. Smith, claiming that Friese-Green had infringed his patents, took him to court. A lengthy court case ensued and eventually, upon appeal, Friese-Green was found not guilty of patent infringement. With continuing financial troubles Friese-Green stumbled on, in a rather noble manner, with his work.

On May 5th 1921 he attended a meeting of filmmakers and distributors debating whether Britain should make her own films or use ready-made ones imported from America. After much discussion is was agreed that film production should continue to be funded in Britain. But before the meeting ended 'there was a commotion round Friese-Greene's seat: people rushed in his direction, but he had collapsed and died. Eighteen pence was all that was found in his pocket. Friese-Greene was the epitome of the struggling inventor who beat a path for others to follow.'[6]

Not until now has the significance of Matthews' association with Fournier D'Albe been fully realised. It was through his work with D'Albe that he became interested in talkies and the simultaneous recording of sound and images onto film. Their association first started back in 1914 when they worked on *Dawn* but continued for a number of years. D'Albe had, back in 1923, been the first person to use wireless to transmit a photograph and, collaborating with Matthews, had for some time been working on a rival television system to that of the legendary inventor John Logie Baird. A description of his apparatus was given in *The Graphic* where

> the image of the speaker, passing through a lens, is directed to revolving perforated grids which produce audio frequencies. Sent by an ordinary wireless transmitter, and picked up by the receiver, the waves pass through a loud-speaker to resonators, and, a powerful light being directed upon the silvered ends of the resonator tubes, the sound waves again become light waves, which next pass to a screen in the form of minute dots which make the picture.[7]

D'Albe patented his system in January 1924, just days before Baird was to give the world's first demonstration of television. The main difference between his and Baird's system was that D'Albe's worked by transmitting the whole picture at the same time with each spot that made up the entire picture sent on a different wavelength or frequency,[8] a system which would ultimately prove impractical. He did give a private demonstration of his television apparatus in Kingston-on-Thames on April 17th 1924 where 'only rough outlines of moving images were transmitted'.[9] D'Albe's system was more complicated than Baird's, which worked by dividing an image into a series of minute sections of light and dark and sending them in rapid succession. Baird would have been well aware of the work being done on a rival television system by D'Albe and Matthews.

The lives of Baird and Matthews have many interesting parallels and although they had several mutual friends, including the

British physicist Sir Oliver Lodge, there is no record of them having ever met or worked together. Born in Scotland on August 14th 1888, Baird demonstrated the world's first television apparatus on the January 14th 1924 in Hastings, East Sussex.[10] His mechanical system, although very successful, would later be dropped by the BBC in favour of an electronic system developed by EMI-Marconi. A well-researched and fascinating account of his life can be read in *John Logie Baird. A Life* by Anthony Camm and the inventor's son, Malcolm Baird.[11] Dogged by repeated bouts of pneumonia throughout his life Baird was an ingenious inventor. At the age of 12 he constructed his own telephone exchange linking up his parents' home with that of four school friends 'living at least two hundred yards away'.[12] Like Matthews, as a boy he improvised his own apparatus from 'household nails wrapped round with wire to make magnets'.[13] His university education was interrupted by the First World War, but rejected as unfit for service he worked as assistant mains engineer with the Clyde Valley Electrical Power Company. The job was so ruinous to his perpetual poor health that he left to start a variety of small businesses including selling soap and 'The Baird Undersock',[14] a medicated sock designed to keep feet warm and dry. His forays in commerce brought him some considerable financial success. After a spell making jam in the West Indies, where he went to recover his health, he returned to the seaside resort of Hastings in Sussex. There he constructed a rather 'Heath Robinson', but brilliant, television apparatus from amongst other things bits of wood, bicycle lenses, biscuit tins and hatboxes. 'The contraption grew and filled my bedroom ... Wireless valves and transformers and neon lamps appeared and at last to my great joy I was able to show the shadow of a little cross transmitted over a few feet.'[15] He later went on to transmit pictures across the Atlantic to America. His work also included colour television, 3-D television, fibre optics, facsimile transmission and radar detection. Late to receive official recognition much of Baird's work would remain classified until long after his death, in relative poverty, in Bexhill-on-Sea on June 14th 1946.

Matthews, after being introduced to the moving image through his work with D'Albe, set about recording sound onto film and embarked on a series of experiments in the ballroom of the New Passage Hotel, where he still had a workshop in the grounds of the hotel. First of all the technical problem of converting sound waves into light waves had to be solved. Aided by the ever-enthusiastic Lynes, Matthews and his assistants began experimenting with a 10 amp arc lamp, a transformer, a selenium cell and a microphone. The lamp projected a beam of light onto a small mirror which reflected the light on to a parabolic mirror which, in turn, focused the light, that was about the size of a coin, onto a selenium cell. With everything in place a gramophone record was played. Sound waves from the gramophone were changed into electrical energy by a microphone which was connected to the arc lamp thereby converting the sound into light. The light from the lamp was projected, via the arrangement of mirrors, onto the selenium cell, which was connected to a loudspeaker placed in a far-off room in the hotel, via a telephone relay. Matthews discovered that the sound of the music could be interrupted by placing his hand in the path of the projected light. The distant loudspeaker reproduced the sound of the gramophone record, much to the consternation of some of the hotel guests. The experiment was a success and convinced Matthews that sound could be turned into light and then back into sound. Having successfully converted sound into light the next step was to record the light onto a light-sensitive celluloid film, that is, to create the optical sound track. After months of work Matthews and his assistants designed and built a beautiful camera that was able to record images and create an optical soundtrack. This was no small achievement and a remarkable testimony to his technical ingenuity. With a mechanism to wind a reel of film at three feet per second the camera had a small slit aperture on the side. The beam of light coming from the lamp, connected to the microphone, went through the slit and was focused onto the moving reel of film where it was recorded. So throughout the entire process, sound was converted into electricity, by a microphone, the electrical energy was then converted into light, which was recorded onto film: a chemical process. The appearance of the sound waves when recorded onto the film was similar to the 'irregular edge of a saw'.[16]

What Matthews wanted to do was record both images and

sound onto a standard reel of film, used by the film industry, thereby making his process commercially more attractive by avoiding the need for specialised equipment. However this presented two distinct problems. The first problem was that action photographs would only leave a small width of the film available to record the sound. The second problem was that the standard speed for filming at the time was sixteen pictures per second which resulted in the recorded sound overlapping and running on into the next frame so that very quickly the sound wouldn't match up with the image when replayed. Matthews' camera solved both problems by using a mechanism that would wind the film more quickly, at a rate of twenty-four pictures per second, but avoided tearing the fragile celluloid film. He managed to achieve a constant and steady winding of the film by adding a flywheel to the camera's winding mechanism. 'Often when it appeared that I was upon the very threshold of success I would find, on running the film through, failure, abject and lamentable. But, as so often is the case, from each failure I gathered a little knowledge, and at last came success.'[17] With sound and picture now synchronised onto the film the next problem was to separate the image from the recorded sound. A device called a 'gate' enabled the two parts of the film, the picture and the optical soundtrack, to be screened from each other on the reel of film. The gate was designed to be fully adjustable thereby allowing the sound track to occupy a narrower band on the film leaving more room from the picture to be recorded – and more room meant greater detail could be recorded.

With the excitement of their triumphant experiments at the New Passage Hotel Matthews wanted a more practical laboratory space and he, along with his small team of assistants, moved all his equipment and apparatus to a new, larger, laboratory at number 2 Harewood Place, Hanover Square, just off Oxford Circus, in London at a cost of £250 per annum. He also employed more technical staff including a Mr Robert Ruddock to work on refining and developing his camera. To demonstrate the practicalities of his camera he set up a small recording studio on the roof of his Harewood laboratory which also had a flat on the top floor. Known as the 'cottage on the roof'[18] it was furnished like a summer house, with creepers and flowers, and was the location for the recording of one of the world's first talking pictures. The number of distinguished visitors calling in on Matthews at Harewood Place grew as his reputation spread and his name would have been mention in the corridors and smoking lounges at both the Royal Institution and the Royal Societies Club. Amongst the influential people who visited his recording studio were: Colonel Bromhead, head of the Gaumont Corporation, a French film company; Lord Howard de Walden, who wrote plays; Prince George, later Duke of Kent; Sir Harry Lauder, a popular entertainer and music hall artist, and various members of universities and laboratories seeking out how he had solved the technical problems of simultaneous sound and image recording.

On September 16th 1921 Sir Ernest Shackleton, the Antarctic explorer, called in at Harewood Place to keep an appointment he had with Matthews. That was the year Shackleton was to embark on what was to be his last expedition to the Antarctic, for he died of a heart attack whilst on board his ship *The Quest*. Entering his laboratory Matthews was alarmed at Sir Ernest's appearance as he looked particularly tired and rather sickly. Quickly taking his chair he told Matthews that it 'was his twelfth appointment that morning'.[19] Matthews asked about his forthcoming expedition and Sir Ernest replied 'I am leaving in a few hours. I'm not feeling too fit, but I'll soon shake that off when I get to sea. Life during the past few weeks has been one long hectic rush – meetings, conferences, and more meetings. I've had very little sleep, but I'll make that up when I get aboard.'[20] Because of his hectic schedule Sir Ernest didn't have much time to complete the interview and Matthews, along with Mr Kingston, his assistant, quickly set up the microphone and camera and the recording was made:

> Oh, well, this is just the start of a new expedition that will certainly circumnavigate the world at a very high altitude. It is the outcome of a dream made possible by the generosity of an old school-fellow – Mr. John Rowett.
> It lies on the knees of the gods at the present moment as to what we shall do. But month by month, as the expedition goes on, our results will be endowed for the benefit of the country and the furtherance of our Geographical Society.

> First of all we go to Cape Town, and then from Cape Town we go through the ice. Shall we get through the ice-floes? Perhaps – well, no one can say. There may be great mountains; there may be plains, barriers or gulfs. But anyhow, three thousand miles of coastline, unexplored, have to be accounted for.
>
> The ship is small, but she is none the less seaworthy. It would be possible to wriggle through the ice. This is a boon, as it would be impossible with a much bigger ship ...[21]

Matthews noticed that Sir Ernest was highly strung throughout the recording but with the interview complete and after wishing the explorer a safe and successful journey Sir Ernest dashed into the busy street to make his next appointment. Sir Ernest Shackleton was later buried in South Georgia on March 5th 1922. After hearing of his death Matthews offered the recording he had made to Sir Ernest's friend, Mr Rowett, who wrote back thanking him for his kind offer and saying that 'I feel you will understand my reasons for not accepting your kind gift'.[22]

News of what was happening at Harewood Place now started to reach the public when on September 28th a reporter for *The Times* wrote:

> The achievements of Mr. Sven Aison Bergland, the Swedish inventor, have been rivalled by those of a British inventor. I have just seen a picture of a pianist giving a recital, and at the same time heard the music, and the synchronization is perfect.
>
> After the war Mr. Grindell Matthews transferred his attention to the subject of the synchronization of photographing object and sound, and he has now completed his work. There is little new in the photographing of sound by means of light rays, for 20 years ago Herr Ruhmer conducted successful experiments. The practical proposition before the modern inventor was how to secure both records upon a single film, in order that the apparatus might be such as to appeal to business men ... The slightest vibrations received by the electrophones cause a tiny mirror of stainless steel to vibrate two pencils of light which are focused on the edge of the film. The film is shown in the ordinary way. There is no trouble about synchronization ... Mr. Matthews' invention may, too, be adapted to other ends. Thus a speech delivered, say, in the Albert Hall, could be reproduced at Colston Hall, Bristol, by means of the ordinary telephone wires, or at as many other places as were desired.[23]

After filing his patent for *'Improvements relating to Photographic Sound Recording'*[24] in November 1922, Matthews was keen to improve the process of photographing sound, seeking to make it more reliable and efficient. One particular problem he had experienced was the tardiness of the electronics caused by using the selenium photoelectric cell. He had experimented by varying the beam of light with a vibrating wire or reed or by adjusting the width of the aperture and the shutter through which the light passed. The camera's variable slit aperture was controlled by an electromagnet enabling the width to be varied. The electromagnet, connected to the microphone, caused the aperture to open and close, in synchronisation, with the sound impulses. The narrow beam of light was focused by the parabolic mirror through the aperture onto the winding film. The 'reed' component vibrated in front of the aperture allowing the beam of light to pass through the aperture and strike the reel of film where it left a zigzag line that varied in width. This is known as variable amplitude. Matthews found that he could replace the reed, which was susceptible to a dampening effect, with a very fine wire, but this had the drawback of being less sensitive and very susceptible to dust. Matthews was to solve all the problems caused by using a wire by inventing a device called 'The Gloson'. This was a tube, filled with a gas mixture of helium, neon and argon and controlled 'the quiverings of light – the frequency up to 6000 a second – in step with electrical vibrations set up in a microphone by the voice, music, or any other sound, so that they could be photographed'.[25] The Gloson emitted a variable light and therefore totally did away with the need for a variable aperture. It had to be manufactured from a special type of glass and Matthews had to employ his own glass-blower to make such a one-off specialised piece of apparatus. The Gloson, it was later discovered, also had medical applications and was used for treating tuberculosis of the eye.[26] In July 1926 Matthews patented the Gloson in both America and Britain as the *'Electric Gas-discharge Tube for Photographic Sound Recording'*.[27]

Matthews enjoyed working at Harewood Place and his days spent there 'were among the happiest of his life'.[28] He would often work continuously through the night, meals were regularly left uneaten, and totally oblivious to the time, when returning the next morning his assistants would find him crouched over his workbench still absorbed in some technicality. Often to be found sleeping on a workbench was 'Twinkie', Matthews' pet black cat. But during the early Twenties Matthews had achieved his goal of recording sound and images on the same reel of standard celluloid film. All that remained now was to sell it to the film industry.

Not until 1927, five years after filing the patent for his camera, did the first talking film appear in Britain. *Blackmail*, directed by the legendary Alfred Hitchcock, was 'the first full length all talkie film made in Great Britain'.[29] So what happened in that five-year period? Matthews thought, reasonably enough, that there would be little difficulty in marketing his camera: after all it had solved the problem of synchronising sound and pictures onto a standard reel of film, he held the patents, he had solved all the technical problems and he didn't have to deal with government bureaucrats. He wrote to Oswald Stoll, whom he knew from the wireless duel at the London Hippodrome back in 1910, who wrote back saying;

> My advice, if such it can be termed, is that yours is a wonderful idea, and eminently practical, but I doubt its commercial value. I don't think this generation will want pictures to talk. I may be wrong and out of date. Anyway, I'm glad to have met you, for I was most interested. Don't think I'm against you.[30]

Time and again Matthews was told what a wonderful idea he had and how practical his camera was, but no film company would take it up or sponsor him. He tried to interest Lord Beaverbrook, the newspaper magnate, who was 'impressed' and thought 'that there were possibilities in the invention'[31] but all to no avail. Matthews was nonplussed and sought the opinion of influential heads of the film industry who told him that taking up his 'fantastic talking picture would mean scrapping millions of pounds worth of capital which we have sunk in production'.[32] Many popular film stars of the day couldn't speak properly and the reason they enjoyed the huge success they did was simply because 'they photographed well',[33] and film companies would not be prepared to write off the millions of pounds invested in their box office stars. Matthews reasoned that surely the cinema going public 'deserve something better than mere synthetic acting'[34] and appealed to them to 'try just one talking picture as an experiment'.[35] But film producers were simply not prepared 'to scrap the machinery fitted to cinemas, or kill the box-office value of stars'[36] and Matthews couldn't convince them otherwise. He recorded that 'it is ridiculous to believe that this generation does not want films and records of the world's greatest orators: film records of grand opera, plays, and drama by the world's greatest artists'.[37] He even gave exhibitions of his talkies to the film trade and other interested parties but he just wasn't able to sell the concept to them.

During the autumn of 1924 Matthews made the first of many trips to America. He first travelled to America to see if he could interest the big American film magnates in his invention and get away from the huge media interest he was the subject of at the time back in England regarding his 'death ray'. It was whilst crossing the Atlantic that he met Jesse L. Lasky, head of the Paramount organisation, who went on to become a very successful film producer. Matthews, overcoming his shyness, seized the opportunity to introduce himself and started talking about his camera and the difficulty he was having in getting the British film industry interested in talkies. But Lasky was just as indifferent, telling him that the public did not want films that talked. Hollywood, it seemed, was quite happy with the status quo and its current domination of the silent film industry. Vsevolod Pudovkin, the Russian film director, said 'there is no artistic future for the talkie. The American method of using sound adds nothing to motion pictures.'[38] The popular American actor Douglas Fairbanks, well known for his swashbuckling roles as Robin Hood and Zorro, made his position clear when he declared 'Speaking films would limit my field of acting, since I depend greatly on acrobatics to put my films over. There is no room for speech in films such as my own.'[39] Unable to sell his talkies or finance and produce his own talking films on a commercial scale, Matthews' patents lapsed. Nobody was interested, Matthews was simply ahead of his time. Why could nobody see the future of cinema?

The advent of talkies was inevitable and their popularity steadily grew during the latter part of the 1920s, with Matthews observing, 'I realised that the film in which the characters speak as easily and naturally as in a stage play was an inevitable development of the motion picture industry.'[40] The last silent film ever made was *Kliou the Tiger* in 1936, directed by Henri de la Falaise.[41]

Had Matthews met Sam Warner on that Atlantic crossing, in 1924, things could have been so very different. Sam, along with his brothers Albert, Jack and Harry, formed Warner Brothers, the well-known American film company that popularised cinema throughout America. A 'self taught genius in mechanics [who] instantly recognized the ground breaking potential' of talkies,[42] Sam Warner persuaded his brothers to adopt the 'Vitaphone' system of sound recording which was the world's first commercial sound recording system. Matthews would later be employed by Warner Brothers for a year between May 1929 and May 1930, as technical consultant for $3,000. During this period, when the inventor was living in America, talkies really started to take off back in England, but by then America had a virtual monopoly on the industry. The increasing popularity of talkies, that Matthews had foreseen years before, forced the fledgling British film industry to accept the technology. Film magnates recalled an inventor who had approached them some years previously talking about some such device that recorded sound onto film. 'Now what was his name ...? Grindell something.'

The film critic G.A. Atkinson well remembered Matthews and wrote an article to the *Daily Express*. He explained how the photo-electric cell,

> the basis of all talking systems in which sound-waves are photographed as light waves, is said to be the invention of Mr. T.W. Case, formerly in the employ of Dr. Lee de Forest whose Phonofilms were demonstrated in London three or four years ago. But a similar device is claimed to have been used by a British inventor, Mr. Grindell Matthews, some years before ... The whereabouts of Mr. Grindell Matthews is at present unknown, but it is probable that he is in New York, and efforts are being made by British motion pictures to find him, as his special knowledge would be valuable at this juncture. If the result of this action, in which there are many millions of money at stake, should show that the Case patent really belongs to Britain, the benefit to this country's talking film prospect would be enormous.[43]

Atkinson did, apparently, manage to contact Matthews, who simply told him that he was under contract to Warner Brothers and due to lack of money and disinterest at that time had allowed his patents to lapse. Between 1919 and 1923 Matthews obtained two full and 23 provisional patents relating to the photographing of sound and movement on the same film strip. This work had cost him about £10,000, of which he managed to get between £7,000 and £8,000 from interested parties. Despite giving demonstrations of his invention to the film trade he failed to attract any interest or further financial backing and because of this all his patents lapsed.[44]

Matthews, although a pioneer of talkies, wasn't the first inventor to have the idea of using light to record sound onto film. Lee de Forest, aware of Matthews' work with talkies, was a prolific American inventor and his story can be read in *Lee de Forest and the Fatherhood of Radio* by James Hijiya.[45] In 1906 De Forest invented the Audion vacuum tube which amplified weak radio signals making them more audible and was an important innovation in the field of electronics. Edwin Armstrong later improved the device making it a practical component for radio devices. The Audion, later renamed the triode, was an improvement of Flemming's diode vacuum tube detector. In January 1920, when Matthews was working at the New Passage Hotel, De Forest, working in the same field, successfully converted sound to light and then back into sound.[46] Later, in July of the following year, just two months before Matthews recorded Ernest Shackleton's farewell speech, he made an experimental talkie with his 'Photophone' and later in November of 1922 he started the De Forest Phonofilm Corporation to make and market his films. The formation of this company would ultimately lead to De Forest having more commercial success than Matthews, who was no businessman, with the marketing of his recording system. De Forest also formed what would ultimately be a doomed partnership with Theodore Case. Case was an inventor of no mean repute, holding many patents of his own, and his photoelectric cell was used in De Forest's Photophone. Collaborating with Case, De Forest made further improvements to his apparatus and on April 15th 1923 he showed clips of his own talkies to the American public at the Ravioli Theatre on Broadway.[47] Between

1923 and 1927 De Forest made several talkies of various celebrities including, in 1924, President Coolidge making a speech. After perfecting his Photophone De Forest, like Matthews, thought he wouldn't have any problems getting the film industry to accept his invention. But again, with echoes of what Matthews had experienced back in England, he had little success. In the summer months of 1924 De Forest, running short of funds, recorded in his journal that he was 'almost out of money'.[48] Then in 1925 came the split ending De Forest's collaboration with Case, who withdrew the licence that granted De Forest Phonofilm permission to use his inventions. His fortunes continued to take a plunge when Warner Brothers in New York started using the Vitaphone system for talkies. The world's first commercial sound system, Vitaphone, manufactured by Western Electric, used a phonograph disc that was synchronised with the running film. The electric signals detected by the microphone were transferred to a 17-inch disc that ran in synchronisation with the film cameras.[49] Warner Brothers used the Vitaphone system to screen John Barrymore in *Don Juan* in 1926. *Don Juan*, although having no spoken dialogue, did have sound effects and music.[50] The world's first feature film with spoken words was *The Jazz Singer* (1927) with Al Jolson saying the immortal words 'You ain't seen nothin' yet.' The film was a huge hit and went and on to enjoy considerable commercial success.

All this happened in America five years after Matthews pioneered the technology back in England. Meanwhile De Forest sold the rights to his Phonofilm to Fox, a rival of Warner Brothers, in October 1926 for $100,000. However Fox had also bought the rights to a talking picture system from Theodore Case, who had continued his research after splitting with De Forest. Case had made further technological progress and sold his system to Fox in July 1926. The newly formed Fox-Case corporation called its sound-on-film system 'Movietone' and it was used to make its first motion picture in January 1927. Fox-Case would later abandon silent films in March 1929. The Fox Movietone system was very similar to Matthews' system of sound recording. Electrical signals from the microphone were detected by a photoelectric cell, converting the sound vibrations into varying intensities of light which reacted with the light sensitive film creating an optical soundtrack.[51] De Forest filed lawsuits against Case for infringement on his patents and eventually received $60,000 in an out-of-court settlement. As all this was unfolding, Matthews had the role of onlooker and was living permanently in America with his second wife, Olive Waite. His work on talkies had cost in the region of £10,000[52] and the film industry wasn't interested in it – well not until his patents had lapsed, but by then it was too late.

Notes

1. www.nationalmediamuseum.org.uk/General/MuseumNews.asp?NewsID=74 (accessed 05/12/07).
2. Brown, G.I., *The Guinness History of Inventions* (Guinness Publishing, 1996), p. 168.
3. Allister, Ray, *Friese-Greene. Close-up of an Inventor* (Marsland Publications, 1948).
4. *The Magic Box* (1951), starring Robert Donat. Directed by John Boulting.
5. www.victorian-cinema.net/rudge.htm (accessed 25/05/07).
6. McArthur, T. and Waddell, P., *The Secret Life of John Logie Baird* (Hutchinson, 1986), p. 116.
7. *The Graphic*, January 26th 1924.
8. McArthur and Waddell, p. 136.
9. Ibid., p. 60.
10. Ibid., p. 34.
11. Kamm, Antony and Baird, Malcolm, *John Logie Baird. A Life* (National Museum of Scotland Publishing, 2002).
12. McArthur and Waddell, p. 68.
13. Ibid., p. 20.
14. Ibid., p. 68.
15. Baird, John Logie, *Television and Me. The Memoirs of John Logie Baird* (Mercatpress, 2004), p. 45.
16. Barwell, p. 78.
17. 'He invented talkies – but too soon', *Daily Express*, June 25th 1935. Swansea County Archive D/DZ 346/11.
18. Barwell, p. 76.
19. Ibid.
20. Ibid., p. 77.
21. Ibid.
22. Ibid.

23. *The Times*, September 28th 1921.
24. http://gb.espacenet.com/ patent number GB206908 (accessed 22/05/07).
25. Barwell, p. 78.
26. Ibid., p. 79.
27. http://gb.espacenet.com/ patent number US1575701 (accessed 22/05/07).
28. Barwell, p. 79.
29. *Cinema Year by Year. 1894–2004* (Dorling Kindersley, 2004).
30. Barwell, p. 81.
31. Ibid., p. 82.
32. Ibid., p. 83.
33. Ibid.
34. Ibid.
35. Ibid.
36. Ibid.
37. 'He invented talkies – but too soon'.
38. Ibid.
39. Barwell, p. 84.
40. 'He invented talkies – but too soon'.
41. www.imdb.com/title/tt0129196/#comment (accessed 21/05/07).
42. www2.warnerbros.com/main/company_info/company_info.html?frompage=wb_homepage (accessed 21/05/07).
43. Barwell, p. 87.
44. National Archives BT 226/4670.
45. Hijiya, James A., *Lee de Forest and the Fatherhood of Radio* (Associated University Presses, Inc., 1992).
46. Ibid., p. 102.
47. Ibid., p. 103.
48. Ibid., p. 106.
49. *Cinema Year by Year. 1894–2004*.
50. Ibid.
51. Ibid.
52. Ibid.

CHAPTER 6

'Prometheus' and the 'Death Ray'

IN 1924 KEEN cinema-goers were watching *The Death Ray*, advertised as 'The Most Startling and Breath-Taking Motion Picture Ever Made!'[1] Shown at several London cinemas including The Tivoli in the Strand and The Majestic in Tottenham Court Road,[2] it brought Matthews an enormous amount of publicity both in Britain, and later in America. The showing of the film also brought him some much-needed capital. His experiments with talkies had cost something in the region of £10,000 and his 'death ray' £1,500. An excellent review of *The Death Ray* can be read on the internet movie database IMDb.[3] For 24 minutes the audience would be astounded as they watched Matthews demonstrating his 'death ray' in all its terrifying glory. The short film was written and directed by Gaston Quiribet and made, not as a piece of entertainment, but to seriously promote and demonstrate the capabilities of the invention. During the film Matthews 'bears no resemblance to a crackpot or "mad" scientist ... He appears ... slim, handsome, dignified – wearing a spotless lab coat and spats ... His laboratory looks like a genuine place of scientific research ...'[4] In the film Matthews and his assistants operate the 'death ray' to kill a rat in a cage, illuminate a light bulb that has no discernible wires, stop a bicycle wheel from spinning and towards the end of the documentary ignite a tray of gunpowder 'which explodes with a satisfyingly large cloud of smoke'.[5]

But the author of the review believes that Matthews was a fake and a fraudster stating that the director, Gaston Quiribet, had worked on movies in the past, movies that were noted for their special effects, and that the film was 'meant to deceive'.[6] The film is, unfortunately, no longer available, recorded onto nitrate film, a medium which is notoriously unstable and deteriorates quickly. However there are some fascinating stills from the film held in the Swansea Archives.[7] Full of atmosphere, they do seem to be quite theatrical, showing Matthews and his assistants in a variety of poses with the 'death ray'. But this wasn't the only film of the 'death ray'. A few years later, around 1927, Matthews made a short newsreel item for Pathé and this can still be viewed on the Pathé website.[8] Introduced as *War's Latest Terror!* Matthews is again trying to convince us of the possibility of his 'death ray'. A smartly dressed and relaxed Matthews appears in the front of his house posing for the camera and can be seen to mouth 'Is that alright?' Then in the next scene a bespectacled, floppy-haired assistant places what appears to be gunpowder on to the top of a wooden box, set up in front of a brick wall. After a few brief moments the gunpowder explodes in a puff of smoke, the result, one assumes, of being subjected to the 'death ray'. A curious collection of devices including a small black box, a swinging pendulum and an electric fan is then shown to the astounded audience. After a few seconds a small spotlight shines onto the pendulum which promptly stops swinging. Then, aimed at the electric fan, the 'death ray' stops the motion of the whirling. As interesting and intriguing as the footage is one is somehow left feeling unconvinced. Why? There is opportunity for Matthews to pull the wool over the eyes of the audience and the more sceptical would say this was exactly what he was doing. When presented with something that, at face value, is unbelievable, it naturally follows that people are sceptical. But that doesn't mean that what they are witnessing is somehow fraudulent.

It would be the 'death ray' that would have a profound and lasting effect on Matthews' career and, rightly or wrongly, define his reputation as an inventor in the eyes of the public. It captured the public's imagination more than anything else he had done previously, with newspapers all around the world reporting details of this terrifying new weapon that could 'end all wars'. It was true drama with fantastic claims, counter claims, international espionage, government intrigue, lies and shadowy figures offering fabulous sums of money for it. The 'death ray' first entered the public's imagination when a masterpiece of science fiction, *The War of the Worlds* by H.G. Wells, appeared on bookshelves in 1898. Wells vividly describes how after landing on Horsell Common, in Surrey, the Martians began their invasion of Earth:

> It is still a matter of wonder how the Martians are able to slay men so swiftly and so silently. Many think that in some way they are able to generate an intense heat in a chamber of practically absolute non-conductivity. This intense heat they project in a parallel beam against any object they choose by means of a polished mirror of unknown composition, much as a parabolic mirror of a lighthouse projects a beam of light. But no one has absolutely proved these details. However it is done, it is certain that a beam of heat is the essence of the matter. Heat and invisible instead of visible light. Whatever is combustible flashes into flames at its touch. Lead runs to water; it softens iron, cracks and melts glass, and when it falls upon water, immediately that explodes into steam ...[9]

A 'death ray' is an indispensable prop to any science fiction story, with heroes and villains appearing on the front of weekly science fiction comics wielding a ray gun in their attempts to destroy their arch enemies. Ming the Merciless used a purple death ray to safeguard his dictatorial throne over the planet Mongo, and the megalomaniac Bond villain Auric Gold Finger tries to slice 007 in half with a red laser. Dan Dare, 'The Biggles of Space', always had his trusty ray gun and Captain James T. Kirk would often be heard telling his crew to 'set phasers to stun!' During the 30s, 40s and 50s stories and articles about death ray technology would regularly appear in popular science magazines. Even today rumours persist and there are stories about such technology with claims the American forces have a 'Heat Beam Weapon'.[10]

But its history stretches as far back as 250 BC when the Greek philosopher and scientist Archimedes invented the solar ray, a huge giant concave mirror to focus the rays of the sun onto invading Roman warships causing them to catch fire. Ever since then it has evolved into an increasingly more sophisticated weapon that has always been linked to wholesale destruction, the product of an evil mind with a mad scientist seen clinging to the apparatus, laughing hysterically, eyes glaring, laboratory coat flapping and proclaiming his hunger for power and control. It's this public image of such a device that explains why Matthews was ultimately remembered as a latter-day Prometheus, a cunning and devious figure, stealing fire from the Gods only to hand it to mortals, with history calling him 'Death Ray Matthews'.

He never liked the title 'death ray' simply calling it what it was, an 'electric beam'. His interest in the 'death ray' was initially sparked when he read newspaper reports that aircraft were being mysteriously brought down over Germany 'by an unknown method'.[11] Numerous articles were appearing in the press describing how 'French planes had been forced down on Germany territory by some mysterious ray'.[12] Matthews doubted the existence of such a 'mysterious ray' at the time, for he noticed that all the unexplained downing of aircraft seem to occur within the close proximity of high-powered radio stations, and assumed that there must have been a connection. But he wasn't the only one taking an interest: scientists from the War Office were also taking a closer look at this phenomenon. They already knew that radio waves could carry electrical energy and considered the possibility that an aircraft's magneto was being shorted out when it flew close to the high-powered radio-transmitting aerial. The magneto is a device, made up of two coils of wire, known as windings, wound closely together, that generates the electricity to create a spark required to ignite the fuel in internal combustion engines. If the electrical current 'leaks' between the two coils then it will short-circuit thereby preventing the spark plugs from sparking and causing the engine to stall. Those scientists thought that the radio waves were causing the magneto to short-circuit and conducted some experiments at a radio station near the Firth of Forth in Scotland which had a Marconi Directional Beam aerial. This type of aerial can send radio waves in a specific direction as opposed to a standard aerial that radiates radio waves like the ripples in a pond. The test they did convinced that small team of scientists that there was a distinct possibility that the action of a magneto could be disrupted owing to an increase in electrical potential or voltage applied to it via radio waves. Their experiments showed that when subjected to shortwave radio waves the fuses and armatures of a magneto would burn out.[13] Radio waves, they discovered, could be used to send, or carry, electrical energy without the need for wires and so by focusing a radio beam onto an engine and varying the wavelength to cover the range of wavelengths of the sparking systems 'the result would be that one sparking plug after another would burn out and the motor would be brought to a standstill'.[14] However the power required to achieve this was very high, but they were convinced that this was the explanation for those French aircraft going down over Germany.

In 1921 The Department of Scientific and Industrial Research (DSIR) had intelligence reports, from the Secret Intelligence Services, of experiments that were rumoured to be taking place in Moscow and Germany. The reports suggested that AEG, the German-based electrical giant that manufactured various electrical equipment, was also involved and was building a machine to project high-energy radio waves. Concerned, the DSIR referred the matter to 'several eminent scientists'.[15] But upon closer investigation the threat was dismissed when the DSIR concluded that 'though experiments of the above nature were in progress, they had probably not yet emerged from the laboratory stage'.[16] They were in fact, as far as they were concerned, at the same stage as their own scientists working in Scotland. However to ensure that electromagnetic waves could not be used to explode ammunition dumps, a scientist from the Admiralty was asked to arrange series of experiments to be carried out in co-operation with the War Office.

Meanwhile Matthews had been doing some work in this direction for himself and was convinced, just like those research scientists in Scotland, that it was possible to transmit energy without wires and conceived the idea of an 'electric beam', which was a 'means of transmitting energy by directive methods for the purpose of interfering with the working of magnetos'.[17] Convinced that he could make a workable device he drew up some plans and started a research programme. The experiments he proposed involved using enormous amounts of voltage and so lots of precautions had to be taken. The wooden floor of his

Harewood laboratory was lined with thick linoleum and the transformers, used to step up the voltage as the electricity surged through the apparatus, were submerged in oil and placed in porcelain tanks with much of the equipment mounted on thick rubber footings. During the early days of those first experiments the air was literally charged with electricity whereupon door handles and other metal objects would become electrically charged, telephones would be affected and fuses in the building were frequently blown, much to the chagrin of the office workers on the floors below. But it wasn't long before Matthews and his team achieved some encouraging results. Using a piece of string Matthews suspended an Osglim lamp from the ceiling with two wires, about a yard long, connected to each end of the lamp to act as an 'aerial'. When subjected to the electric beam it 'lit brilliantly at the first attempt'.[18] He recorded that this could be achieved using 500 watts of power.[19] An ordinary domestic light bulb can use up to 100 watts. Matthews realised that considerable energy was 'leaking' away as it travelled through the air, a problem that he would solve by using a 'carrier beam' of ultraviolet light. Then, with echoes of *War of the Worlds*, he was able to melt plate glass 'with incredible rapidity'.[20] Within a few weeks Matthews and his team were 'able to explode gunpowder, light a lamp and kill vermin at a distance of sixty-four feet',[21] which was the length of his laboratory. Little did those people hurrying past Oxford Circus realise what was happening on the top floor of number 2 Harewood Place. After those first successful experiments they felt confident enough to try something a bit more ambitious and they now wanted to see if they could use their electric beam on a magneto. Acquiring a small petrol engine, they mounted it onto the testing bench and set it running at full revs. Switching on the electric beam, Lynes stood back from the noisy, smoky, engine and Matthews, standing at the controls of the electric beam, aimed it at the all-important magneto. To their utter delight the engine promptly stalled. Lynes examined the magneto and saw that it had indeed been shorted out. All this achieved without the aid of wires over a distance of 64 feet. Matthews noted that the whole thing had been brought about 'by the transmission of electrical energy over a beam directed on to the engine'.[22]

So what exactly was the 'death ray' and how did Matthews claim it worked? It consisted of two main parts: a specialised electrical generator and a carrier beam to act as a conductor and to direct the electrical energy.[23] The secret, Matthews claimed, was in the carrier beam along which the power travelled. Previous attempts by other inventors had been made to project beams of heat waves or electric waves in the same manner as radio waves but Matthews' electric beam worked in a different manner. The carrier beam provided an uninterrupted path along which the power was sent, acting as a conductor of electricity just as an electrical wire does, with high-voltage, low-frequency electrical current flowing along it. It was the electrical power transmitted along the beam that could be used, for instance, to burn the windings within a magneto. However the carrier beam could be used on its own, that is without the electrical power flowing along it. Since this beam acts as a conductor of electricity, should it play against the windings of a magneto it would result in windings merely being short-circuited and not destroyed.

The carrier beam passed through a Fresnel lens in the same manner as in a searchlight. Two electrical currents were then superimposed onto the carrier beam, one of high frequency and one of low frequency, the latter carrying the power. The carrier beam could be made to be invisible or visible, enabling the operator to track a moving target. At the time scientists conjectured that the carrier beam was ultraviolet light, something that Matthews denied. But the general consensus was that ultraviolet light would do the job of a carrier beam. Having a very short wavelength, ultraviolet light will ionise atoms. When an atom, such as the atoms that make up the gases of the air, is unionised, it remains stable and unable to conduct electricity. However when subjected to ultraviolet light an atom will gain a charge and form an ion. Each individual ion gains a positive charge and will thereby conduct electricity.

In an effort to maintain secrecy and minimise speculation, Matthews ordered his equipment and supplies from a variety of different contractors. But in spite of these precautions Fleet Street journalists started hearing rumours that he had invented an 'invisible ray' that would 'stop a motor working, kill plant life, destroy vermin, and explode gunpowder, fire cartridges, and light lamps'.[24] An artist's impression of him at work in his laboratory appeared in the newspapers and offered the public a glimpse into Matthews' laboratory. It appears to be a long, remarkably tidy room, with checkered lino flooring and down one side, running the entire length of the laboratory, a long workbench with various pieces of scientific paraphernalia. A bespectacled Matthews is bending over a small device, which bears a resemblance to a camera, placed on a small table with wires trailing to what one can only assume is the power supply. In the distance is a small petrol engine mounted on a workbench, the magneto of which is the target for the 'electric beam'. Throughout the experiments accidents weren't unheard of with, on one occasion, Lynes – after having accidentally stood in the path of the beam – being 'knocked out'.[25] Matthews himself was to injure his eye and would later, whilst in America, have to have surgery to save his eyesight.

He realised though that to be of any real practical value, such as using it to bring down enemy bombers for instance, the distance over which the energy could be transmitted had to be increased far beyond the limited reached within his London laboratory. The confined space of his laboratory meant that it wasn't possible to use the necessary energy to achieve this and so he set about planning long-range trials which involved handling far greater amounts of electricity. At that time Matthews was, just like the reports that British Intelligence had received about scientists in Germany, only at the laboratory stage and he now wanted to take it much further. But long-range trials presented a new set of technical problems. He had to generate much higher voltages and a method of transmitting and focusing the energy over a much greater distance, and he desperately needed funding. Those unexplained reports of aircraft being downed over Germany still lingered in the columns of the press and the *Daily Mail* journalist F.W. Memory decided to investigate further. Memory would later, in 1933, be part of a team sent to Scotland to investigate claims of the existence of the Loch Ness Monster. Interviewing two well-known scientists – Sir Oliver Lodge and Sir Archibald Low, whom Matthews would meet in later life – about the unexplained downing of aircraft, Memory wasn't convinced by their answers, and smelling a story he decided to call in on a British inventor he had heard of, working in London. Remarkably, Memory was allowed access to Matthews' laboratory, was given a guided tour and shown what was going on, with the understanding that what he saw and heard was 'off the record'[26] and that he would not let anything go to press. This was a remarkably risky move on Matthews' part. Did Matthews simply want to allay any fears regarding what he was doing, was he being naive, or did he realise that a leak would be inevitable and the resulting publicity would help further his cause when he would, as he inevitably knew, have to deal with the government whom he didn't trust in the slightest? Both the inventor and the journalist discussed the reports of 'unexplained' downing of aircraft and Matthews explained his theory, that he thought radio waves had something to do with it, and his current attempts to transmit power through the air without the use of wires. Despite such a risky step taken by Matthews, Memory kept his promise not to go to the press. However over the next few weeks public interest continued to grow with the rumours fanning the flames and with the publication in that same year, 1924, of Fournier D'Albe's book *The Moon Element* interest in what Matthews was doing was to mushroom. With a detailed account of *Dawn*, the £25,000 payment, his work for the British Government on wireless control systems and the success he had managed to achieve with his inventions during the First World War, Matthews became a celebrity, the press descended on his Harewood laboratory and the media circus began. The 'death ray' was now well and truly public property. Here, thought the public, was a scientist known to have carried out work for the government in the past, working on a 'death ray': a device that some claimed could wipe out entire armies and bring an end to wars.

Headlines like 'Death Ray Matthews' started to appear in the newspapers all over London. Matthews was shocked and alarmed by what he read. Dozens of articles full of inaccurate information and fantastic claims of what the deadly 'death ray' could and would do in the wrong hands. With the lack of any technical details the imagination of journalists simply made good the deficiency. All the false claims and misleading newspaper articles were to give Matthews a notoriety that was to dog him for the rest of his life. With huge media interest and wild claims of what Matthews was up to he inevitably came to the attention of the British Government, who already had intelligence reports that the Germans had developed something that was capable of stopping an aeroplane engine. Having done some experiments for themselves, the authorities were well aware that the rumours could be true. Matthews received a letter from Wing Commander Bowen[27] of the Air Ministry, inviting him to demonstrate his electric beam. He was surprised to receive such an invitation and was very reluctant to give a demonstration because he new that his electric beam was just not ready: it was, despite what the newspapers were saying, *'little more then a peashooter'* (author's italics).[28] This invitation is all the more surprising since documents held at the National Archives[29] show that government officials had, in light of their previous dealings with him, been 'ordered not to correspond or have any dealings with Matthews'. But in view of intelligence reports and what Matthews was rumoured to be up to, the government were obviously having second thoughts about their decision to ignore him, with memos finding their way into the 'in' trays on the desks of government officials suggesting that they 'discuss the invention of Mr. Grindell-Matthews'.[30]

Matthews knew it was too early to give an effective demonstration with repeatable and reliable results and that any attempt at a demonstration now would undermine his work in the eyes of the government. Getting no reply, Wing Commander Bowen telephoned Matthews and invited him to come and see him at the Air Ministry to talk the matter over. So early in March 1924 Matthews and Captain Edwards, his solicitor, visited the Air Ministry and spoke to both Wing Commander Bowen and Major Lefroy. During the meeting Matthews told them that he would give a demonstration when he felt that he was ready to do so, saying that he was 'busy experimenting in his laboratory'[31] but 'hoped to arrange to go down to the country where facilities would be available for long-distance tests'.[32] He went on to explain that he was trying to get financial backing and that, presently, he had failed to do so. Despite the media frenzy, he wanted to continue his experiments in private 'until such time that I should be able to perform long distance tests'.[33] Matthews was adamant: it was too early, and that all he had, at that time, was a small-scale working model. But Bowen wasn't to be so easily dismissed: if Matthews had something then it was up to him to make sure the government had a piece of the action, and he wrote to the inventor again asking him to give a preliminary demonstration in light of the fact that he thought a full demonstration was premature and that 'facilities would be given for the open air and longer distances'.[34] The government was certainly keen to see what he had, which was rather curious, when they had been so dismissive of his earlier work. This second letter presented Matthews with a dilemma. He still remembered the Marconi Scandal and knew the ministers could act in a cloak-and-dagger fashion and simply didn't trust them. But in light of the government's second offer, another refusal would show he had a lack of faith in his work and his credibility might be undermined. Besides, a proper demonstration would prove that the claims in the press were wildly exaggerated. After talking it over with Lynes and Edwards Matthews decided to go ahead, took his courage in both hands and made arrangements for an official demonstration. After much bureaucracy on the part of the government, arrangements were made for Matthews to give three separate demonstrations on the same day to the Air Ministry, the Admiralty and the War Department. Initially only the Air Ministry was to be present but when the other departments got wind of the demonstration they wanted to see it. The demonstrations were to take place at Matthews' Harewood Laboratory on May 26th 1924.[35] A gathering of academics, civil servants and officials including Major-General Thuillier, Colonel Craig, Professor Andrade, Mr Smith, Mr Pickles, Wing Commander Bowen and Major Wimperis all witnessed the 'death ray' in action on that momentous day. It would be Major Wimperis, the director of scientific research for the Air Ministry who would later ask Robert Watson-Watt, the pioneer of radar, to look into the feasibility of a 'death ray'.

Matthews and his team spent a feverish couple of weeks preparing for the demonstrations: everything had to be perfect, no slip-ups, and nothing short of a convincing demonstration would do. They managed to get everything ready and on the morning of May 26th a calm and confident Matthews welcomed the party of visiting officials and lead them up the three flights of stairs into his well-equipped laboratory. He showed them around his laboratory, told them the nature of his work, explained what they were about to see and that there were other interested parties that were keen to have his electric beam, including a French firm with whom he was currently in negotiations. Matthews also explained that he, along with a small group of investors, was presently setting up a company, 'G.M. Company Ltd.', to finance the project. With authorised capital of £40,000 in 40,000 £1 shares the vendors held 20,000 shares, with the remaining 20,000 shares offered for private subscription with an agreement that Matthews would retain 51% ownership of his 'electric beam'.[36] The syndicate was headed by Mr A.H. Caley, a wealthy London businessman with Mr E.G. Gubbins and Mr J. Sanbourne providing finance to enable Matthews to continue with his work. Beginning the demonstration, Lynes handed Major Wimperis an Osglim lamp, asking him to hold the terminal while Lynes held the other end. Matthews disappeared into a cabin situated at the far end of the laboratory and was able to switch the lamp on and off 'at will'.[37] But Wimperis thought the whole thing very unconvincing, reporting that lighting a lamp under such circumstances was a simple matter and that he was 'rather surprised to find that the inventor should imagine that one would be impressed'.[38] The inventor then invited the officials over to examine a small single cylinder petrol engine mounted on to the top of the workbench. Starting the little engine, Wimperis was asked to control the throttle and when he wanted the engine to stop he was to raise his hand and Matthews would use his electric beam to cause the engine to stall. Raising his hand 'the engine stopped altogether and had to be restarted'. Wimperis said 'I examined the engine and took out the sparking plug. It was quite an ordinary plug and the general connections of the engine fastened onto the bench were all apparently quite normal.'[39] But Wimperis remained unconvinced and rather suspicious:

> one of the attendants was standing near the engine, and I could not see the other two while watching the engine. Such examina-

tion as I made showed nothing amiss and *had that engine experiment been carried out in an Air Ministry installation the inventor would have proved his case*'. [Author's italics.] Carried out in his own laboratory one cannot say. *I could see nothing underhand about it*, [Author's italics] but it would be easy to achieve the same experiment if carried out on one's own bench.[40]

Mr Smith from the Admiralty described the ray as blue in colour with a slight tinge of red. This would back up the fact that an ultraviolet carrier beam was used. Matthews demonstrated that his beam could act through sheet metal by placing a piece of tin plate directly in front of the engine's magneto. Smith wanted to see if there were any electrical connections between the motor and the bench and asked Matthews if the engine could be moved: 'He did not like this suggestion, and explained further that he was in a great hurry.'[41] Officials thought this was a strange thing for him to say but he did have to perform the demonstration twice more that same day. It was important that the engine remain in position for it was difficult to precisely focus the beam and any movement of the engine made a demonstration all the more difficult. However one of the assistants did move the engine to the floor and Smith noticed 'that there were two brass bolts through the table and nuts to help keep the motor down' but he didn't see 'whether there were any leads connected with the bench'.[42] Professor Andrade was far more cynical and remarked that 'there was every opportunity for concealed wires, or coils underneath the bench' and that the demonstration was 'worthless'.[43]

Later that same day the officials, from each division of the armed services, who witnessed the demonstrations held a conference at the War Office where Major-General Thuillier told those gathered that he had no authority to offer the inventor any money until he performed further tests in a government laboratory or with an engine supplied by the government in his own laboratory. Major Wimperis and J.B. Bowen interviewed Mr W. Stocken who had financed Matthews back in 1908. Stocken told them that he thought he knew how the 'death ray' worked but had been advised by the Air Ministry to keep the information to himself and that no discussion with the Air Ministry had ever taken place over the matter.[44] After some deliberation Wimperis wrote to Matthews suggesting that the demonstration be repeated with a motor-cycle engine supplied and placed in situ by government technicians. Should, Wimperis went onto explain, the test be successful, the inventor was to be awarded the sum of £1,000. The condition of this offer however was that he was not allowed enter into negotiations or reveal any details of the invention to a third party for a period of 14 days. Such a proviso suggested that the government were certainly taking Matthews seriously and were obviously concerned about his negotiations with the French firm that he had mentioned. But he refused their offer on the grounds that the French firm had made him an offer based on exactly the same demonstration the government had witnessed, and he could not afford to wait 14 days while dithering bureaucrats ruminated over their indecision. He told them he was off to France. Worried that Matthews was now going to the Continent, Wimperis called in at his laboratory. Speaking about his move to France he told Major Wimperis that he would not reveal any particulars of how his ray worked to his French associates. He was going to France because the offer that had been made to him was very attractive, in financial terms, and given his present financial situation he couldn't refuse. Wimperis was anxious that despite Matthews' assurance that he wouldn't reveal any details of how his invention worked, he might be pressed into doing so and pointed out to the inventor that he was able to resist such pressure by stating that he was held by a certain act of Parliament that prevented him from divulging details of inventions connected with the armament to a foreign country without the sanction of the British Government.[45] He went on to suggest that when he returned from France they arrange another demonstration, '*not* [to] *investigate how he carries out his side of the performance, but merely to watch the effect on the engine to satisfy ourselves that no extraneous agency was at work*' (author's italics).[44]

So who was the French firm that Matthews was dealing with? The British Government wanted to know; they had good reasons to keep tabs on him and made some covert enquiries. They were in fact compiling a dossier on Matthews which could not be inspected 'without special authority'.[47] The Intelligence Services discovered that it was a firm called Chantiers du Rhone, based in Lyon, owned by a French engineer called Eugene Royer. The Air Ministry had already written to the British Embassy in Paris to enquire if the French authorities were aware that the Germans were in possession of a 'Ray' or 'Anti-aircraft Waves' and had received information that the French and German authorities were known to be in possession of an apparatus from which rays could be projected to a great height, causing aeroplane engines to break down.

The British Consulate in Paris, Mr Charles Dick, wrote to the Air Ministry informing them that Matthews was, indeed by the end of May 1924, in Lyon with Captain Edwards, his solicitor. The authorities were obviously watching Matthews very closely and Mr Dick told the British Intelligence that Matthews had stated in the press that he was in France negotiating an agreement with Mr Eugene Royer, an engineer and inventor who had a steel foundry where he was doing experiments in connection with an invention for cutting steel plates with oxyacetylene. Royer did himself hold patents relating to implements for cutting materials. The following year, 1925, he was granted a patent for a 'Flame cutter intended to function in water' and he was granted two more patents in 1926 for a novel '*Oxyacetylene Blow Pipe*' and another for a '*Firing device for cutting torches.*'[48] It's plain to see why he would be interested in Matthews' electric beam as perhaps he thought he could use it as another type of powerful cutting tool. Matthews had previously demonstrated it could melt plate glass. Mr Dick went so far as to say that reliable sources informed him that Mr Royer 'was a most unreliable person, in a financial position bordering on bankruptcy'.[49] So worried were the British Government of the 'death ray falling into the hands of a foreign power'[50] that they advised the British Consulate to contact Matthews and warn him about the dangers of dealing with Mr Royer given his financial situation. But Royer had impressed Matthews, who considered the foundry to be a good place to continue his work with his electric beam. There was a large electrical power plant which Matthews wanted and the foundry would be able to supply him with the necessary facilities to manufacture the parts he required for the large-scale version of his electric beam: 'M.Royer, the proprietor of Chantiers du Rhone, and myself are merging our interest into the company, *except with regard to the "beam" and my other war inventions, which remain entirely in my hands to be dealt with as I consider best*' (author's italics).[51]

The deal with Chantiers du Rhone included a £10,000 cash payment and a profit-sharing scheme with Matthews holding a position as a director with a handsome salary.[52] It is hardly surprising he turned down the offer from the British Government when he was being offered such attractive financial terms. It is interesting to note that Matthews claimed that Chantiers du Rhone were only interested in a 'portion of the profits which may be derived from the sale of my electric beam *or from other dealings concerning it*' (author's italics).[53] Was he sending a guarded warning to the British Government? Matthews went on to say that 'In addition to the French offer, which I have accepted, I have received offers from other countries – notably from the United States and *from Germany, which country is particularly keen to claim my invention*' (author's italics).[54] Determined as Matthews was to work with Royer, the consulate warned the Air Ministry to be very cautious in dealing with him should he become in any way connected with Royer. However shady the government thought Royer was there are documents showing that on October 20th 1924 Royer filed for Patent No. 606 260 '*Projection à distance Phénomenès invisibles de haute fréquency électrique*'[55] or the 'remote projection of invisible high frequency electricity' in other words, the 'death ray'. Surely this gives Matthews and Royer credibility in the fact that they had a working electric beam. But the fact that he did patent it and that details were held in the French Patent Office suggest that it wasn't a secret or even classified; he hadn't sold it to the French Government. Matthews had formed a partnership with Royer, who patented the electric beam. The patent describes how the device is able to transmit power to stop electrical machines, and consists of, amongst other things, a projecting mirror and a transmitting sphere able to create a conducting magnetic field. The high-frequency waves emitted from the sphere are reflected by a mirror and projected to the target along a conducting magnetic field thus allowing the 'transport of force, destruction of the effect of the generators, transformation of the molecular state of the matter, change of colour of the bodies, etc.'[56] So here was the 'death ray'. It's interesting that Royer applied for and was granted the patent; Matthews' name doesn't appear on the document. He had given, or rather sold, his electric beam to his French colleague who, no doubt, had his own plans for it.

Whilst in Paris the press interest became so intense that Matthews found it difficult to stay at any hotel without being mobbed and was forced to rent a private house to get some degree of privacy and avoid the media. Everywhere he went there was a media scrum. But it wasn't just government officials who were anxious about recent events: his financial backers – Caley, Gubbins and Sanbourne – had some questions of their own. Having got wind that Matthews was on his way to France they wanted to protect their interests in the electric beam and they quickly obtained an injunction, issued by Mr Justice Greer, from the courts preventing him from taking it abroad. Clutching the injunction in his hand Gubbins and his solicitor sped to Croydon airport in an attempt to intercept Matthews. But they were too late, missing him by a matter of minutes; they watched helplessly as his aeroplane took off. Only moments before Matthews had posed for photographs as he climbed into the aircraft: photographs that would appear in the daily newspapers under the headlines 'Death Ray Secret Lost to England!' the following day. The publicity would only bring him more support from the British public and possibly force the government's hand into offering him a better deal over the 'death ray'. Landing at Le Bourget Aerodrome in Paris a short while later, Matthews announced, to the ever-present press, that he was in no way obliged to his backers back in England. They had, according to a report in the *Star*, May 28th 1924, 'an option but it expired six weeks ago'.[57] However eventually a settlement was arranged with Matthews and his backers that meant the 'death ray' would remain in England. The settlement involved the formation of a company (G.M. Company Ltd) that acquired the 'death ray' on terms that Matthews found 'quite satisfactory'.[58] Chantiers du Rhone was to continue to maintain an interest in his other inventions but also have a financial interest in the syndicate formed to keep the 'death ray' in England. The *Westminster Gazette*, May 30th 1924, reported that a contract had been signed on May 17th at the Hotel Petrograd in Paris whereby a percentage from the sales of the 'death ray' would go to Chantiers du Rhone.

Members of Parliament had watched all this in a state of consternation. A quick look at the newspapers dropping through their letterboxes did nothing to reassure them and questions were asked in the House of Commons on May 22nd 1924: only four days before government officials had called in at Harewood Place to see the 'death ray' in action. Mr Leach MP, Under-Secretary of State for Air, said 'My attention has been called to the claims made for the invention referred to, the Air Ministry is in touch with Mr. Grindell Matthews. But for the present it is inadvisable to make any full statement upon the matter.'[59] Asked if the British Government had been offered Mr Grindell Matthews' invention, Mr Leach replied 'I cannot answer that.'[60] So the government decided that stonewalling was the best way forward. The matter wasn't to end there for on May 26th, the same day as Matthews had given that official demonstration, because of the media storm the Under-Secretary of State had to make a statement to Parliament: 'Mr. Grindell Matthews was offered and refused an opportunity to demonstrate his invention under conditions which would satisfy either scientists or businessmen' and the inventor was offered 'every facility to give a demonstration under conditions satisfactory to himself and to the Services'. But the conditions weren't offered on a basis that Matthews found satisfactory *to himself.* Mr Leach went on to say that

> The Departments have been placed in a difficult position in dealing with Mr. Grindell Matthews, partly because of the Press campaign which has been conducted on behalf of this gentleman and partly because this is *not the first occasion on which the inventor has put forward schemes for which extravagant claims have been made.* [Author's italics.] As a result, the Departments are unable to accept Mr. Grindell Matthews' statement without scrutiny which, apparently, he is not prepared to face.[61]

When asked if Matthews had been paid £25,000 by the Admiralty 'for an invention to direct vessels by wireless, and were the Admiralty exactly throwing money away?' Mr Leach replied that he 'couldn't answer for another department'.[62] But it is now known that Matthews was indeed awarded the money for this work. So why not acknowledge the fact that Matthews had done work for the government in the past and been paid for it? Were they trying to undermine his credibility? Lieutenant Commander Kenworthy MP asked if the government was 'quite certain that there is no value in this invention, and, if so, are they taking any steps to prevent it going into other hands outside this country?' Mr Leach said that as the government had not been allowed to test the device 'we are not in a position to pass judgment on the value of this ray'.[63] Mr Thorne MP queried as to why 'according to the newspapers, some steps have been taken to prevent this individual [Mr. Matthews] selling this particular invention to France?' Mr Leach said that 'romantic minded public is inclined to take too much notice of the Press campaign.'[64] Viscount Curzon made a claim that 'the Air Force had an invention which will do all that Mr. Grindell Matthews was able to do in the course of his experiments'.[65] Mr Leach replied that 'every phenomenon produced by Mr. Grindell Matthews can be readily reproduced by the people in our Department, but that does not say there is anything of value in the phenomenon'.[65] So here Mr Leach is acknowledging that Matthews' device is capable of doing something having previously remarked that they hadn't been able to test his device. Mr T. Johnston MP wanted assurance 'that no public money will be spent on any blackmailing individual who threatens otherwise to sell his invention to a foreign power'.[67] Was Matthews trying to blackmail the government? On being asked if the government had issued the injunction to Matthews prior to him going abroad, Mr Leach told the House that 'No, we had nothing to do with that matter at all.'[68] Lieut. Colonel Watts-Morgan ended this debate in Parliament on more humorous note: 'would it be possible to introduce this invention to perform an operation on the Opposition?'

The matter still wasn't allowed to rest and in the following month on June 4th Mr Maxton MP raised the issue of the safety of government technicians whose task it was to examine such devices offered to the nation and sought assurances that they 'would not be used in reckless experiments of this description'.[69] He wanted to know how one particular official from the Ministry had, during a recent demonstration, placed himself within ten yards of the ray which was supposed to 'destroy men and machines at a distance of several miles'.[70] He didn't know anything as *a veil of secrecy having been drawn over the proceedings*, [author's italics] but I am disposed to think that, if the claims of the inventor were anywhere near being justified, this man must have been destroyed, and his body, perhaps, disposed of in some evil way'.[71] Matthews had never claimed at any point the his electric beam as it was could kill a man: it wasn't anywhere near powerful enough and he had told them that what he had was 'little more than a peashooter'. Matthews wanted funding and support to scale up his work making it a practical answer to the threat of enemy aircraft. His work, like that of scientists on the Continent, 'wasn't beyond the laboratory stage' and he was desperate to get it to the battlefront. In laboratory tests Matthews had transmitted energy as far as his laboratory space had allowed: 64 feet. However he was struggling to achieve directive control, ultraviolet light being invisible, and he was engaged in finding a way that he could both transmit and see the beam of light that would act as a conductor of the electricity. In the confined space in which he was working he was unable to achieve this.

Mr Leach told the concerned MP that indeed an official had 'placed himself within the path of the ray as part of the test to which he desired to be subjected' but he assured his honorable friend that 'he was doing well, and, when I last saw him, he displayed no evidence of any injury'.[72] What Mr Leach did not tell the house was that under the watchful eye of Matthews when the official moved too close to the ray he switched off the power and so there was no reason for the official to be anything other than perfectly fine. Admiral Kerr was far more eager to acquire the ray for the British Government. He showed the device to 'a surgeon who was quick to detect its possibilities as a therapeutic agent'.[73] Kerr 'got in touch with many wealthy people, trying to get them to purchase the secret for the country.[74] Kerr, an advocate of Matthews and his work, was the former Deputy Chief of the Air Staff, and was in no doubt that the 'ray is a very real and terrible thing ... if Mr. Grindell Matthews' new invention is good enough for France, it is good enough for us'.[75] He had tried to interest various organisations and individuals in Matthews' work but all to no avail. After Matthews' death Kerr wrote to his biographer, Ernest Barwell, saying that 'Grindell Matthews was a truly wonderful man, and of great value to his country. If he had been listened to more seriously and quickly we would have gone a great way ahead of other nations, but, unfortunately, we are always very slow at taking up new ideas.'[76] The *Standard* reported that Matthews was in negotiations with Sir Samuel Instone, Chairman of S. Instone & Co., who said of the inventor: 'Matthews is an inventor of repute and we [are] aware that he has not been well treated by the Air Ministry.'[77] Sir Samuel and his brother Theodore were wealthy entrepreneurs with business interests in shipping and aviation. Sir Samuel is reported to have held one of the world's first telephone conversations with an aircraft in flight in August 1920. He would have had a keen interest in Matthews' work in wireless communication, particularly his successful attempt at radio communication with an aeroplane in flight back in September 1911.

But whilst all this was going on Matthews still managed to spend time in his laboratory and it wasn't long before he had solved the technical problems encountered in trying to extend the distance over which his electric beam worked. Matthews was able to do some long-range tests on the tiny Island of Flat Holm, five miles off the coast of Swansea in the Bristol Channel. Back in May 1897 Marconi and his colleague William Preece had transmitted a Morse code message by wireless from Flat Holm to Lavernock point on the south coast of Wales.[78] The full-sized electric beam made an impressive sight. A large structure shaped like a drum, mounted on top of what looked like a beehive that was made of porcelain and rubber that insulated the 'gun' from the ground and contained the induction motor. On the outside of the drum were three fixtures that looked like megaphones, and housed coils of wire. These coils would glow red hot as thousands of volts surged through them but the arrangement allowed the heat to dissipate thereby preventing the whole thing from overheating. Inside the drum, which could be rotated through 360 degrees, was the actual mechanism, powered by 50,000 watts, a thousand times more powerful than the tested prototype, which projected the carrier beam through the Fresnel lens into the air.[79]

Meanwhile the Secretary of State visited the distinguished scientist Sir Ernest Rutherford in his rooms at Cambridge University. Considered to be the 'Father of Nuclear Physics', Sir Ernest had been a student of J.J. Thompson whilst studying at Cambridge University. It was Thompson who had officiated at the test of *Dawn* back in October 1915. Thompson cast doubt on that test of *Dawn* saying that the late Lord Fisher was 'mainly instrumental in Mr. Grindell-Matthews obtaining a Government grant of £25,000 during the war'.[80] Over a quiet sherry the Secretary told Sir Ernest about the grilling he had recently had in the House over the wretched 'death ray' and the crackpot scientist claiming to have invented it. 'Could,' asked the Secretary, 'Matthews really have a working "death ray"?' 'Sir Ernest Rutherford was very outspoken about the death ray and knew all about the episode of Lord Fisher. He had no doubt at all that the "ray" would prove to all intents and purposes worthless'[81] and that 'Grindell-Matthews is best avoided.'[82]

It is ironic that the press, although championing this lone inventor's cause, were, by making extravagant claims, undermining any credibility that Matthews was seeking. When compared to the sensationalist articles his device was, initially, as he claimed 'a peashooter', so naturally the government became suspicious. The *Sunday Express* invited Matthews to give his own account of what he had been doing at Harewood Place and on June 1st 1924 an article entitled 'My Ray' by H. Grindell-Matthews appeared in the *Sunday Express* newspaper.[83] In this article Matthews acknowledges that his electric beam 'has been the subject of so many conflicting accounts in the Press'. He had been researching and 'devising a means for transmitting energy by directive methods for the purpose of interfering with the working of magnetos, and even with the possibility in view of setting fire to the plane itself'. Confirming that tests within the laboratory had enabled him and his team to 'cause a magneto to cease functioning by the transmission of electrical energy over a beam directed on to the engine,' he went onto explain that, at that time, he was trying to gain financial backing so that he could undertake long-distance tests that he was planning. Acknowledging that the government had expressed an interest in his beam as early as December 1923, he said he was keen that his country 'should have the benefit'.

But the prevailing air of misleading facts and fantastic claims forced Matthews to write to the Press Association.[84] He mentions the changes that the government kept requesting: first only one department, the Air Ministry, was to witness his invention working, but then the Admiralty and the War Office wanted to be there. Acquiescing to their demands they then weren't satisfied with the trial they had all witnessed and demanded a second one with the proviso that should they find it acceptable he wasn't to speak to any third party about it for two weeks. Matthews went on to explain the reason for refusing to give a second trial: he wasn't prepared to use an engine supplied and installed by government technicians. The press interest was unrelenting and Matthews would fight his corner. He wrote a letter to newspapers giving his version of events:

5th June, 1924.

Sir,
One gets tired of correcting the inaccuracies and misstatements of the representatives of the Government.

In reply to Mr. Maxton, who asked whether a Government expert had stood within ten yards of the so-called Death Ray which was supposed to kill at a distance of some miles, and escaped unhurt, Mr. Leach facetiously replied that this was so, and the official was doing well.

Even to the meanest intelligence I should have thought it was apparent that the Ray, unless directly focused on the object of its purpose, must be capable of control and be innocuous to those operating it.

It was naturally understood that the Government representatives should not be subjected to any risk whatever, nor was more than a small fractional part of the power available used. Should, however, they be desirous of testing its death-dealing properties, I should be pleased to oblige them as soon as I am established in my laboratory, but this of course, must be at their own risk. I will take no responsibility.

Yours, etc., H. Grindell Matthews.[85]

According to Barwell 'the challenge brought no reply'.[86]

With this letter appearing in the Press the appetite for information grew and the public, liking a good drama, were demanding more information. The press was, to its credit, always on his side and willing to help fight the inventor's corner. Extracts from an article that appeared in the *Daily Mail* reported that:

> A very grave mistake has been made by the British Imperial Staff in not thoroughly investigating Mr. Grindell Matthews' invention. The facts which have from time to time appeared in our columns show that there is a strong prima facie case for testing his ray and bearing the cost of his experiments. If it is such a weapon as is supposed, the danger of its loss to us as a nation may be very great indeed...[87]

With the prevailing climate of the interwar years the press interest was understandable. People wanted to believe that Matthews had a device that could make aircraft fall from the sky and stop future wars and were angry as to why the government had treated him in such a dismissive manner. Memories of the First World War were still in people's minds.

Matthews wasn't the only one working on a 'death ray' and there are records showing that other inventors threw their hats, or rather their lab coats, into the ring. Notable amongst those in England was Dr. T.F. Wall of Sheffield University. He was, according to the *Leeds Mercury* of May 29th 1924, working on a device 'believed to be almost identical with that of Mr. Harry Grindell Matthews'. His work had led him to discover that '300 volts would be sufficient to destroy life' but he went onto explain that it 'does not necessarily follow that the same arrangement will be suitable for the interfering with the magnetos of engines'.[88] Dr Wall had intended his research to be for the benefit of his country but said that any dealings with the government 'should be fair on both sides'.[89] A fairness that had always eluded Matthews. Over in America an inventor called Bernard Johnson claimed to have built the 'Z-Ray',[90] a rival device to the 'death ray'. Dr Charles P. Steinmetz, a German electrical engineer, also working in America, reported that he had made 'Human Lightning Stroke'.[91] Mr Scott, from San Francisco, was reported to have offered a device similar to the 'artificial thunderbolt' to President Coolidge for $250,000.[92]

Many inventors, ranging from the serious to the positively eccentric, were all staking a claim on the 'death ray'. But perhaps the best remembered inventor having a go at this was Nikola Tesla. A thorough analysis of his life and work can be read in *The Life and Times of Nikola Tesla – Biography of a Genius*.[93] Tesla's relationship with his 'death ray' is unerringly similar to that of Matthews and his device. Matthews has been referred to as 'The Welsh Tesla', such is the similarity of these two brilliant inventors. Tesla was indeed one of the most remarkable inventors of the twentieth century, and with roughly 700 patents to his name he is widely considered to be 'The Father of Modern Electrical Technology'. Born in Croatia in 1856, he studied electrical engineering in both Austria and Prague. He worked for the Continental Edison Company in Paris in 1882 before emigrating to America in 1884. It was whilst working in Paris that Tesla built an induction motor. This type of motor is driven by Alternating Current (AC) and is more versatile than a Direct Current (DC) motor. After arriving in New York Tesla worked for Thomas Edison but their working relationship was a difficult one and they soon parted company. The animosity between them prevented them from sharing a Nobel Prize for Physics despite their enormous contributions to science and technology. Tesla was a strong advocate of the AC system of electrical distribution which he saw as far superior to the DC system favoured by Edison. AC travels much further and more easily than DC electricity and once Tesla had invented AC generators, induction motors and other electrical components that could run on AC electricity, the fate of the DC system was sealed. One of his better-known inventions is the Tesla Coil, a device that steps up voltage and can create impressive electrical arcs. Tesla would demonstrate his coil by lighting up fluorescent tubes without any connecting wires during the many lectures he was invited to give around the world. Tesla coils are still an integral component of everyday electronic devices such as radios and televisions. His later work included a wireless world broadcasting tower, the Wardenclyffe Tower, on Long Island, turbines, robotics and wireless communication. Poor at managing his finances, his hotel bills often went unpaid, and he accumulated considerable debts. He was wonderfully eccentric, with fastidious habits, a keen pigeon-fancier and suffered from Obsessive Compulsive Disorder. But it was his more outlandish ideas such as communication with other planets, claims that he could split the earth in half and, like

Matthews, his 'death ray', that resulted in his being universally regarded as a mad scientist. Tesla's 'death ray' worked differently than the one invented by Matthews. He claimed to have invented a 'particle beam weapon'. Just like Matthews, Tesla thought it was possible 'to transmit electrical energy without wires and produce destructive effects at a distance'.[94] Tesla doubted that Matthews could have made a ray in the way he claimed saying that, 'I worked on that idea for many years before my ignorance was dispelled and I became convinced that it could not be realized.'[95] Although both devices were not actually lasers they were considered to be forerunners of the laser. A laser, such as a ruby laser, is made from a crystal of ruby that is shaped into a cylinder with a mirror placed at one end and a partially reflecting mirror at the opposite end. A high intensity light is shone onto the ruby crystal which excites the electrons in the atoms that make up the ruby crystal. The excited electrons give off energy as light with the mirrors at each end of the crystal reflecting light backwards and forwards exciting yet more electrons. A red or ruby light escapes through the partially reflecting mirror as a laser beam. After perfecting his particle beam weapon Tesla, just like Matthews, failed to interest his government.

When he died on January 7th 1943, the FBI and the Office of Alien Property (OAP) took an interest in Tesla's activities and seized all his papers, documents and various pieces of equipment. They were concerned that his work might end up in enemy hands. The papers relating to his work on his 'death ray' were examined by the American Government who appointed Professor Trump to take a close look at them. After going through some of the numerous trunks that had been seized, Professor Trump concluded that 'nothing valuable would be found' and 'that it would be useless to look in the 29-odd trunks which ... had ... been stored since 1933'.[96] However he did retrieve some papers covering work Tesla did in the latter part of his life and passed them on to the OAP. Reporting back to the government Trump said 'upon the basis of my examination, it is my opinion the Tesla papers contain nothing of value for the war effort, and nothing which would be helpful to the enemy if it fell into enemy hands'.[97] Tesla had, in 1937, written an article called 'The New Art of Projecting Concentrated Non-dispersive

Energy through the Natural Media'. This contained the most detailed account of how his particle-beam weapon worked. Trump played down the significance of this article which is, to this day, classified as top secret by the American military.[98] The government also ordered that his Wardenclyffe Tower to be destroyed. Similar events happened upon the death of Matthews, whereupon it has been reported that the British Government seized his papers and stripped his laboratory of equipment.

With the failure of his talkies and the storm of the 'death ray' raging in England, Matthews made the first of many trips to America during the late autumn of 1924, when he embarked on a transatlantic liner to New York. He was relieved to be leaving the media scrum and the resulting fallout of the 'death ray' but was mistaken if he thought he was going to have a quieter life. His reputation as an inventor had preceded him and the 'death ray' saga was front-page news over in the States. Journalists camped outside the Vanderbilt and Waldorf Hotel in New York, where he stayed, all of them clamouring for information. At around that time, Matthews had noticed a marked deterioration in his eyesight. He struggled to read fine print, detailed work was difficult and his eyes tired easily. He would often complain that he could see swirling spots and the pain in his left eye gave him splitting headaches. Lynes had also noticed the deterioration in his eyesight and urged his boss to go to the opticians. Whilst in Manhattan Matthews consulted an eye specialist, Dr Henry Beers, who diagnosed a rare form of Ambylobia, or 'snow blindness' and told him that he would need surgery to prevent him losing the sight in left eye altogether. He underwent surgery immediately. After a lengthy and delicate operation Dr Beers had managed to restore some vision to his badly affected left eye and the sight in his right eye was completely restored.[99] But Matthews would, to his chagrin, have to wear an eye patch at regular intervals over his left eye to prevent any further deterioration in his sight. He hated wearing it: the tight head-band gave him headaches, messed up his hair and made him feel self-conscious.

Matthews thoroughly enjoyed his first visit to America, travelling and touring extensively, and it wasn't long before he found himself falling in love with America and the American people. He found their enthusiastic attitude to inventors and entrepreneurs to be such a stark contrast to stuffy old class-conscious England. It was all so intoxicating. He thought financial backing would be much easier to come by with the gung-ho attitude of American businessmen. He lived, for a short while, on Park Avenue in New York, and Long Island. Whilst in New York, in February 1925,[100] he met Miss Beulah Louise Henry. Born in 1887 in Tennessee, Miss Henry was a remarkable inventor – so much so that she is fondly remembered as 'Lady Edison'. During her lifetime she was granted 49 patents, including a vacuum ice-cream freezer, a spring-limbed doll, and sponges with soap in the middle. One of her more successful inventions was a parasol that had a detachable cloth cover which came in a variety of different colours. Her parasols became extremely popular and she set up a company, 'Henry Umbrella and Parasol Company', to manufacture and sell them.[101] Matthews and Miss Henry formed a close friendship, with a shared passion for cats, and became enthusiastic supporters of each other's work. Their meeting was widely reported in the press, which had the effect of giving Matthews more kudos. In the constant sunshine his health improved and so did his outlook. After spending several months in America his mind was made up: he wanted to move out there permanently and planned to set up a laboratory in New York where he wanted to continue his work and market some of his inventions. He sailed back to England to collect his equipment, settled various matters and in April 1925 found himself sailing back across the Atlantic for America. He didn't sever all ties with England or Europe, however, for over the next six years he would make numerous visits back and forth across the Atlantic on various matters relating to his work.

The showing of the 'death ray' film at theatres such as the Lyceum, in New York, had turned Matthews into a celebrity, and made him considerable sums of money. The public were astonished by the 'death ray'. Matthews declared that he had 'disposed of his invention in America' and 'England has definitely lost the chance of obtaining my invention'[102] but he would not divulge the buyer's price. The buyer could well have been Warner Brothers, who later employed him for a year during

1929/30, for it was reportedly used as a prop on film sets. Unfortunately no records of Matthews working for Warner Brothers remain. The sale of the 'death ray' to the USA was widely reported in the American Press, but the American government denied that there had been an official purchase of the sinister secret.[103] No 'official' purchase? So did they purchase it 'unofficially'?

Matthews had for some time been romancing an American beauty called Olive Waite. Tall, attractive with dark hair and eyes, he found Olive intelligent and utterly captivating from the moment he first met her and quickly fell in love. Olive was, at the time, filing for divorce from her husband, Malcolm Waite, an actor and amateur boxer. Shortly after meeting they decided to marry. They travelled extensively together with Olive accompanying Harry whenever he had to return to England. After her divorce came through they were married, on September 17th 1925, by civil official in Pitlochry, a tiny hamlet in Perthshire, Scotland, described by Olive as 'the most romantic spot in the world'.[104] After getting married they returned to New York, aboard the *Mauretania* and went to Marble Collegiate Church, to solemnise their marital vows in a service conducted by Reverend Dr Murphy. 'Smiling happily, Mrs. Matthews confessed that the lovemaking of her husband had conquered her heart with the same dispatch his brain children won him the recognition of scientists. She added that they will be at home to their friends at Beaverbrook Farm, Middle Neck, Long Island'[105] (where Matthews had a laboratory). He would be financially dependent on Olive during the six years he spent in America – with the exception of the $3,000 he was paid by Warner Brothers between May 27th 1929 and May 27th 1930 he didn't earn any money. But just as with his first wife Katy, the marriage wasn't to last long and they were to divorce in September 1930.

Although not on record it is also possible that Matthews may have called in on the celebrated inventor Nikola Tesla and the writer Hugo Gernsback. Hugo Gernsback was born in Luxembourg but later became an American citizen. Remembered in some circles as 'The Father of Modern Science Fiction',[106] he wrote for numerous science fiction magazines, and was also an amateur wireless enthusiast and inventor with numerous patents to his credit. In *Ralph 124c 41 +*,[107] a science fiction story Gernsback wrote in 1911, he mentions a device called an Actinoscope which was an early radar device. Gernsback and Dr Severinghouse, who was working as a physicist at Columbia University, made an unsuccessful attempt to copy Matthews' ray with heat beams and ultraviolet light.[108]

So did Matthews really have a working 'death ray'? The government certainly witnessed a working model. Although failing to live up to their expectations it did show potential, for why else would they have shown such interest and consternation with it? There are also eyewitness accounts of Matthews using his electric beam during the late 1930s when he was living in Tor Clawdd, his laboratory in the Welsh mountains. Although this was not the 'death ray' as claimed by the media, it certainly supports the theory that he had a working model along the lines that he claimed. The injury he had sustained to his right eye seemed to be genuine and could conceivably have been caused by a device that fired high-velocity particles. Although there is no known reference of them ever working or collaborating together both he and Tesla must have been aware of each other's work for the American press widely reported their work. But just as doubts remain about Matthews ever inventing a 'death ray', such a mystery surrounds Tesla and *his* claims to have invented a 'death ray'.

Did Matthews try and blackmail the British Government into accepting his invention? After all he did have a deep-seated mistrust of them and harboured ill-feeling towards them. He never felt he got the credit he deserved and was possibly trying to raise his profile as a scientist. As a commercial venture the 'death ray' was a flop and he only seems to have antagonised his investors. Eugene Royer, his French colleague, seems to have just vanished, along with his patent. But inventing is an expensive business. Matthews wanted to invent, but had a poor grasp of financial matters, and a penchant for living in hotels and on credit; his investors wanted to see a financial return on their investment and in the end both parties were to be disappointed. But despite the ultimate failure to get his electric beam taken seriously Matthews' idea to use such a device for the defence of cities was really the birth of Star Wars. The idea in itself was visionary and revolutionary and in the 1920s Matthews was having a go at realising such a fantastic idea.

Notes

1. Swansea County Archive D/DZ 346/7.
2. Swansea County Archive D/DZ 346/12.
3. www.imdb.com/title/tt0303894/ (accessed 05/06/07).
4. Ibid.
5. Ibid.
6. Ibid.
7. Swansea County Archive D/DZ 346/18 and D/DZ 346/2.
8. www.britishpathe.com/product_display.php?searchword=grindell+matthews (accessed 10/06/07).
9. Wells, H.G., *The War of the Worlds* (Modern Library Classics, 1996).
10. http://boston.bizjournals.com/boston/stories/2004/11/29/daily30.html (accessed 22/12/07).
11. 'My Ray' by H. Grindell-Matthews, *Sunday Express*, June 1st 1924.
12. Barwell, p. 90.
13. National Archives File Air 5/179.
14. Ibid.
15. Ibid.
16. Ibid.
17. 'My Ray'.
18. Barwell, p. 90.
19. Swansea County Archive D/DZ 346/10: article from *Evening Journal*, New York, July 21st 1924.
20. Barwell, p. 90.
21. Ibid., p. 91.
22. 'My Ray'.
23. Swansea County Archive D/DZ 346/13.
24. 'My Ray'.
25. Barwell, p. 91.
26. Ibid., p. 90.
27. National Archives File Air 5/179.
28. Barwell, p. 90.
29. Ibid.
30. National Archives File Air 5/179.
31. 'My Ray'.
32. Barwell, p. 93.
33. Ibid.
34. Ibid., p. 94.

35. National Archives File Air 5/179.
36. Daily Province, Vancouver, May 30th 1924.
37. National Archives File Air 5/179.
38. Ibid.
39. Ibid.
40. Ibid.
41. Ibid.
42. Ibid.
43. Ibid.
44. Daily Province, Vancouver, May 30th 1924.
45. Ibid.
46. National Archives File Air 5/179.
47. Ibid.
48. http://v3.espacenet.com (accessed 11/12/07).
49. National Archives File Air 5/179.
50. Ibid.
51. 'My Ray'.
52. *New Zealand Times*, May 30th 1924.
53. National Archives File Air 5/179.
54. Ibid.
55. Patent No. 606 260 'Projection à >distance Phénomenès invisibles de haute fréquency électrique', Institute National De La Propriété Industrielle.
56. Ibid.
57. *Star*, May 28th 1924.
58. 'My Ray'.
59. Hansard Government Records, p. 2393, Oral Answers May 22nd 1924.
60. Ibid.
61. 'My Ray'.
62. Hansard Government Records Vol. 174, p. 422, Oral Answers May 28th 1924.
63. Ibid.
64. Ibid.
65. Ibid.
66. Ibid., p. 423, Oral Answers May 28th 1924.
67. Ibid.
68. Ibid.
69. Hansard Government Records Vol. 174, p. 1410, June 4th 1924.
70. Barwell, p. 96.
71. Hansard Government Records Vol. 174, p. 1410, June 4th 1924.
72. Ibid.
73. Barwell, p. 100.
74. Ibid., p. 99.

75. Ibid., p. 97.
76. Ibid., p. 99.
77. Ibid.
78. Weightman, G., *Signor Marconi's Magic Box. How an Amateur Inventor Defied Scientists and Began the Radio Revolution* (HarperCollins, 2003), pp. 25-6.
79. *Frankfurt Journal*, November 6th 1924.
80. National Archives File Air 5/179.
81. Ibid.
82. Drama Documentary, *The Man on the Mountains*.
83. 'My Ray'.
84. National Archives File Air 5/179.
85. Barwell, p. 96.
86. Ibid., p. 97.
87. Ibid.
88. *Leeds Mercury*, May 29th 1924.
89. *Frankfurt Journal*, November 6th 1924.
90. *Waterbury Republican*, November 16th 1924.
91. *Frankfurt Journal*, November 6th 1924.
92. *Champaign Gazette*, Illinois, November 12th 1924.
93. Seifer, Marc J., *The Life and Times of Nikola Tesla. Biography of a Genius* (Citadel Press, 1998).
94. Ibid., p. 387.
95. Ibid., p. 427.
96. Ibid., p. 453.
97. Ibid.
98. Ibid., pp. 454 and 455.
99. *Brooklyn Eagle*, November 26th 1924.
100. *The Boston*, February 18th 1925.
101. www.ieee.org/web/aboutus/history_center/biography/henry-b.html (accessed 26/12/07).
102. *Washington Post*, March 2nd 1925.
103. *Alton Telegram*, March 5th 1925.
104. Swansea County Archive D/DZ 346/11.
105. Ibid.
106. Siegel, Mark Richard, *Hugo Gernsback, Father of Modern Science Fiction: With Essays on Frank Herbert and Bram Stoker* (Borgo Press, 1988).
107. Gernsback, Hugo, *Ralph 124c 41 +* (Modern Electrics, 1911).
108. Seifer, p. 427.

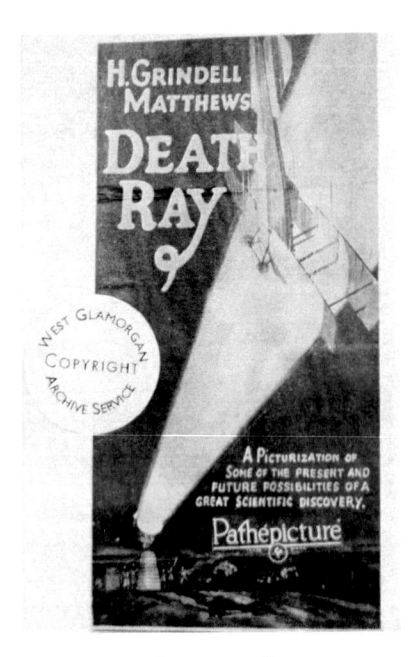

Death Ray Advert.
(Swansea County Archive)

Matthews' vision of travel in the future.

Artist's Drawing of Death Ray.
(Mary Evans Picture library)

Matthews with his talking projector, an invention that the film industry weren't interested in.
(Barwell, 1943)

The first ever camera to record both sound and images onto the same reel of celluloid film. It was this camera that Matthews used to record his interview with Sir Ernest Shackleton.
(Barwell, 1943)

Matthews and his assistants making a talkie, 1921.
(ITN Source)

Colourscope.
(Barwell, 1943)

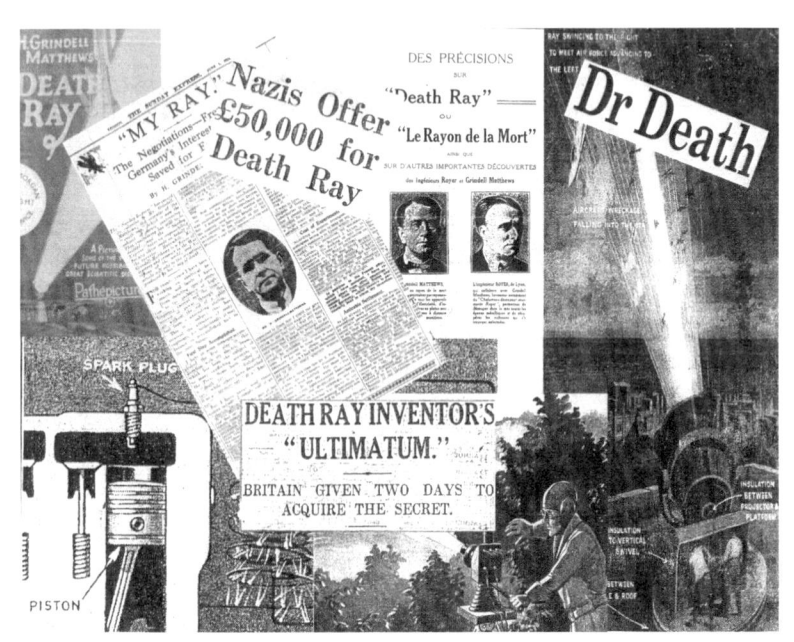

Death Ray Collage.
(Swansea County Archive)

Sky Projector.
(Getty images)

The Keppel's Head Hotel that Matthews and his assistant used as their lodgings.
(www.keppelsheadhotel.co.uk)

Matthews and Captain Edwards, Matthews' solicitor, leaving the Air Ministry after their meeting with Commander Bowen to discuss the 'death ray'.
(Barwell, 1943)

Front page of the French patent filed by Royer for the 'death ray'.

An artist's impression of Matthews working in his Harewood
laboratory.
(Mary Evans Picture Library)

Matthews and the 'death ray'.
(ITN Source)

The 'death ray' in action as shown by Pathé in 1924.
(ITN Source)

Matthews (nearest the camera) operating the 'death ray' in his Harewood laboratory.
(Barwell, 1943)

Matthews and his assistants with a motor bike used to test the action of the 'death ray' on a magneto.
(ITN Source)

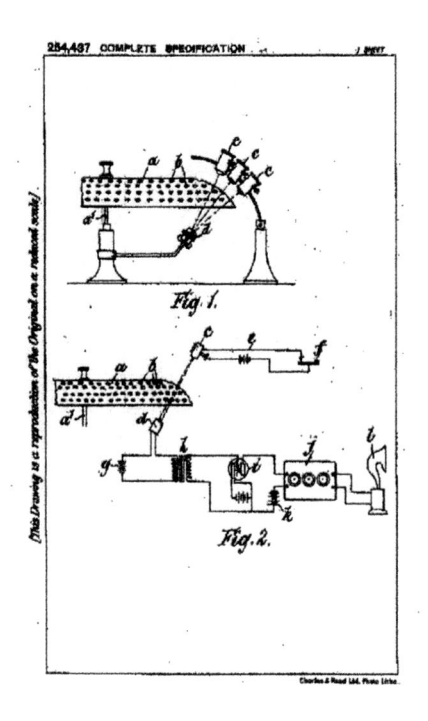

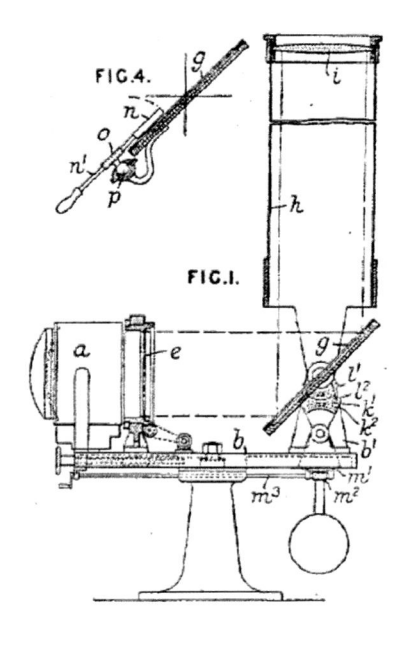

The Tone Wheel, 1925.

An illustration from the Mark I Sky Projector Patent filed by Matthews in 1926.

Matthews playing the Luminaphone.
(Swansea County Archives)

The Sky Projector.
(Barwell, 1943)

Aerial photograph of Tor Clawdd.
(Ioan Richard)

Tor Clawdd as it looks today. Built by Matthews in 1934 and where he spent the last seven years of his life.
(The author)

Rhydypandy Road running past Tor Clawdd. Matthews would use his electric beam to stop passing cars.
(The author)

Tor Clawdd.
(The author)

Tor Clawdd in the 1930s.
(Barwell, 1943)

The Mason's Arms, Clydach. Matthews' local where he would often go for a pint with the locals and where local myth has it he met Winston Churchill to discuss his wartime inventions.
(The author)

The armed forces using the rocket device to deploy the aerial minefield. Matthews never got the recognition for this invention.
(Barwell, 1943)

A rare photograph showing just one of the aerial torpedoes returning to earth after being fired. The trailing wire hanging from the torpedo is serrated enabling it to snag against the leading edge of enemy wingtips.
(Barwell, 1943)

An aircraft on Matthews' private runway behind Tor Clawdd.
(Clydach Historical Society)

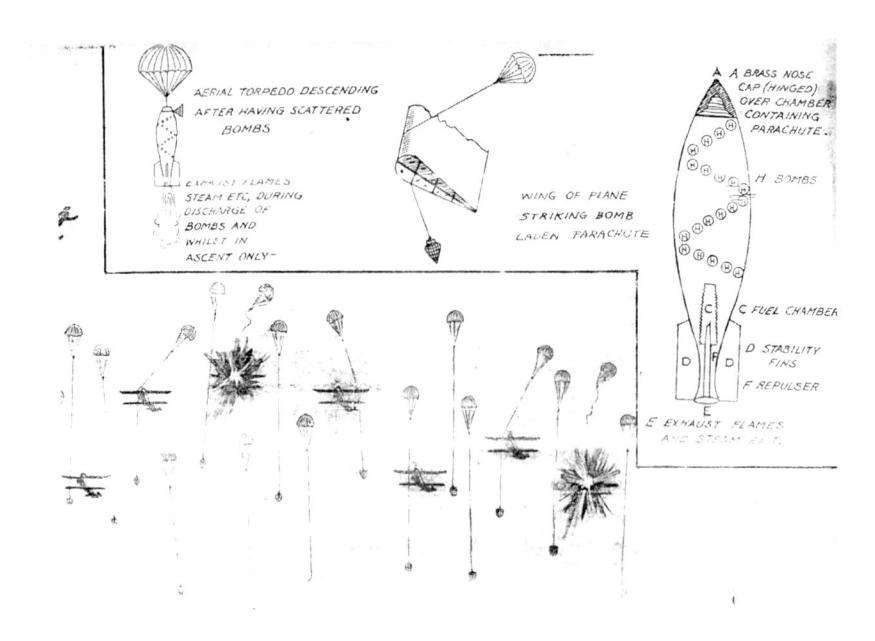

Matthews' plan for his aerial defence scheme to protect cities like London.
(Swansea County Archives)

The 'Flying Flea' Matthews helped to build.
(Helen Macduff)

CHAPTER 7

America and Bankruptcy

WHILST IN AMERICA Matthews was hoping to have some success with his 'Luminaphone'.[1] This was a musical device which was played by the use of light beams and illustrates the diversity of Matthews' work. His Luminaphone was an improvement on an earlier invention of his, 'The Tone-Wheel', that he invented whilst back in London on April 24th 1925.[2] The improvements he made resulted in the Tone Wheel being more compact and able to produce a wider range of musical notes. The patent for the Luminaphone gives his address as West 55th Street, New York and describes how the Luminaphone, a beautiful electronic musical instrument, works: it is played just like a piano, the keys of which operate a series of small lights that shine through a rotating metal disc, with a series of different sized holes, onto light-sensitive cells connected to an amplifier and loud speaker. When the playing keys are pressed a corresponding light is switched on and shines onto the light sensitive cell. Depending on the number and size of the holes on the rotating disc the light is shining through a corresponding musical note is produced via the loudspeaker. This particular invention demonstrates Matthews' mastery of electronics, sound, light and selenium. Whilst in America he also examined the novel concept of converting sound into colour. Barwell claimed that he had designed an instrument called a 'Colorscope' that projected the vibrations of a voice onto a screen and then converted them into colour, with low tones showing up as red and higher tones as violet. With no two voices being exactly the same the Colorscope would show what colour a person's voice was, with each one being as unique as a fingerprint.

The Luminaphone wasn't the only idea Matthews returned to when he was in America. In 1928, whilst in Long Island, he returned to an idea he had first had in 1925: the Sky Projector. The purpose of the Sky Projector was that of advertising, a growing industry back in the late 1920s and early 1930s. Made from a powerful arc lamp, focusing lens, a plane mirror and a stencil, the Sky Projector would project an image, printed on the stencil, high into the night sky without the need for a screen to project the image onto. A very similar device was used to summon the comic hero Batman, who would make his first appearance nearly twelve years after Matthews first patented the idea. Were the creators of Batman, Bob Kane and Bill Finger, inspired by Matthews' Sky Projector? Designed to project an image to a height of up to 15,000 feet, the size of the image was proportional to the height at which it was projected.[3] That is, the higher the image, the smaller it appeared. An unknown American friend of Matthews who worked in the advertising business thought that it had enormous potential as an advertising medium and was prepared to offer financial backing to develop the idea. Enthused, Matthews returned to London to finalise the design and organise the building of a prototype projector. One of the principal elements of the projector was an enormous lens and so he called in on Sir Charles Parsons, at C.A. Parsons and Company in Newcastle. Sir Charles was an engineer best remembered as the inventor of the steam turbine, and had some considerable knowledge and expertise with optical equipment which his company made for searchlights and telescopes. When Matthews put forward his idea for the Sky Projector, Sir Charles failed to be enamoured, telling him that it would be very difficult to construct and, in any case, would be prohibitively expensive not only to make put also to power. But Matthews wasn't so easily discouraged and turned to a manufacturer of glass lenses in Wetzler, near Frankfurt, Germany. Dr Ernest Leitz had a reputation as one of the world's finest manufacturers of lenses which were used in many optical instruments including microscopes and binoculars. Matthews had, in the days he worked on talkies, used a Leitz projector, and thought that the German manufacturer just might be able to produce the sort of specialist lens he was looking for. He found Leitz, who was quite excited at the prospect of making such an enormous lens, to be far more encouraging than Sir Charles; Leitz suggested an immediate trial. Over the next few weeks Matthews and Leitz improvised with what equipment they had and made a prototype of the Sky Projector. Using a 75-amp arc lamp they secured the huge lens, with a stencil, to some scaffolding and, checking that they had the focal lengths just right, they switched on the power supply. As the large arc lamp slowly warmed up the filament began to glow throwing light through the giant lens and a stencil bearing the letters E and L, the initials of his German host. Fortunately that evening the weather was perfect; a clear, crisp night with no cloud cover. Peering intently at the darkening sky Matthews and Leitz were thrilled to see the letters 'E L' appear brightly, high up, in the night sky over the German countryside. Convinced they had an idea that worked, Leitz began to manufacture a lens to Matthews' specific requirements and he returned to America keen to build his Sky Projector.

Upon his return to America, Matthews approached the Sperry Gyroscope Corporation, a major manufacturer of gyroscopic instruments and electronic equipment, based on Long Island, New York, where he had his workshop. Matthews got hold of one of the most powerful lamps available at the time, a Sperry high-intensity arc searchlight that had a 'beam candle power of four hundred million'.[4] Invented in 1915 by Elmer Ambrose Sperry, who founded the company, this high intensity arc lamp was used by the American forces during both world wars.[5] Completely satisfied that the technical problems had been solved and realising that it was indeed possible to build a Sky Projector, Matthews managed to get financial backing from a mysterious figure called Count Valambrosia. Little is known about Valambrosia except that according to files held in the National Archives, Kew[6] he gave the inventor £6,000 for a 40% interest in the Sky Projector. With financial backing from a wealthy patron Matthews ordered two Sky Projectors to be built and patented the invention in Britain, France and Canada.[7] He would later patent an improved version of his Sky Projector when he returned to live in England.

The completed apparatus was large, for it had a focal length of five metres. A parabolic mirror at the end of a tube reflected light from an arc lamp in a horizontal direction, through the stencil, bearing the image to be projected. The horizontal beam was then reflected, by a plane mirror set at a 45° angle, upwards at a right angle along a second tube, through the focusing lens and into the night sky. The second tube was counterweighted and could pivot enabling the direction of the light beam to be altered. The horizontal arrangement of the first tube prevented heat from the powerful searchlamp from damaging the stencil and avoided the problem of hot material dropping from the lamp onto the reflecting parabolic mirror. The orientation of the whole apparatus was adjusted by hand-wound wheels and worm gears, allowing the apparatus to be swivelled and tilted in any direction. Stencils could have movable parts so that the projected image would have the illusion of motion, such as a dog wagging its tail. Parts of the reflecting plane mirror could, it was suggested, have movable segments that again could animate the projected image. The stencil was mounted in a circular carrier which could, by means of a spindle wheel, be slid along the length of the horizontal tube and thereby be brought into focus. The stencil could also be rotated, allowing the image to be seen from different angles. An alternative method involved winding a series of stencils in front of the lamp, from one drum to another drum just like the film in a camera, allowing a series of different images to be projected.[8]

Matthews formed a company called 'Adastra Incorporated' of Albany, New York, in 1928, to exploit the Sky Projector and transferred his shares in the company to his wife, Olive, who had financed him in his work. He said that the Sky Projector could be used for conveying coded messages, advertising and for electioneering. Along with his financial backer, Count Valambrosia, he arranged a demonstration of the Sky Projector for the French car manufacturer Citroën. So impressed were they with what they saw, that Henri Letellier and André Citroën bought the French rights to the Sky Projector.[9] But Citroën weren't the only ones showing an interest in the Sky Projector. The *Daily Mail* entered into negotiations with Matthews and his financial backers to acquire the Sky Projector for £10,000 but, in the end, the negotiations faltered and came to nothing.

Matthews arranged the American debut of his Sky Projector to be made from the roof of the Capitol Theatre on Broadway, using the apparatus to project the words 'Go to the Capitol' onto the side of nearby skyscrapers. Amidst much publicity excited crowds gathered to see the spectacle. One improvement that Matthews had made to the Sky Projector included a colour screen through which the light shone, allowing colour images to be projected.[10] Now a grandmaster when it came to publicity stunts, he and his assistants continued to give spectacular demonstrations of the new Sky Projector, including using the side of the Paramount Theatre to project exciting images. Huge cheering crowds would gather to watch the spectacle of a swooping eagle and the American national flag float in the sky and sweep up and down the side of skyscrapers. But misfortune, never far away from Matthews' side, struck when a gale blew his apparatus off the roof of a building in Brooklyn, New York, where he was giving a demonstration, damaging it beyond repair. Fortunately he had built two models and was able to use the other one.

It was around this time, July 1930, that Matthews decided to move permanently back to England. America had been a mixed experience for him. His contract with Warner Brothers had come to an end. The film studio had employed him as an engineering consultant at a salary of £3,000 during 1930-31. His marriage to Olive was also to end. It had been on the rocks for some while. Matthews was not an easy person to live with and Olive would have had much the same experience as Matthews' first wife Katy. He would be away for long periods of time, making several journeys back to England and Europe on various business matters. His finances were more often than not rather bad and she had financed virtually all his work. A workaholic and rather 'other worldly', he was obsessive about his work and could be moody. When progress on a particular idea was slow he would become depressed. None of this made for a particularly easy marriage. Olive was a socialite, vivacious and outgoing and enjoyed being on the New York social circuit: the sort of things that Matthews, like most inventors, didn't really find appealing.

Matthews returned to England and endeavoured 'to find persons to provide capital to exploit this Sky Projector'.[11] But shortly after his arrival his finances started to go seriously awry.

He had managed to interest some investors into his Sky Projector and a company called 'G.M. Sky Projector' was set up, by an associate of his, Mr Bloom, who himself was to later go bankrupt, to acquire the patent for the Sky Projector and take it to the European market. It was set up with a nominal capital of £300 in various shares to develop the apparatus with the inventor receiving £150 worth of shares. Matthews' older brother, Alfred, also received a number of shares for a loan of £100. Matthews was also appointed a director and consulting engineer and received £310 in fees. Count Valambrosia was given a 40% stake in the company for he still had a financial stake in the Sky Projector. Bloom had financed the company to the extent of approximately £1,000 for which he received shares. Unable to find more money, Bloom sold 1,500 shares to Matthews and the remainder to Col. Orde, for £250, who then became a director. Both Orde and Matthews made a verbal agreement, to pay a further £1,000 to Bloom if the company 'made good'.[12] But the £1,000 was never paid in the end. Matthews argued that Bloom wasn't en-titled to the money because he, Bloom, had 'undertaken to finance the company until the projector was put on the market. Because it never actually made it to market he never got the £1000.'[13] Matthews sold the American rights to his Sky Projector to the company for more shares which he shared out between his wife and Count Valambrosia.

Meanwhile a representative of Wrigleys of Chicago, Mr Malakoff, expressed an interest in the Sky Projector and asked for a demonstration of how the machine worked, which was duly given. Malakoff appeared impressed as the technical details of the apparatus were explained to him. But Matthews later heard that, shortly after, Malakoff had built a machine according to what he had seen and lodged an 'interference' against G.M. Sky Projector Ltd at the patent office. G.M. Sky Projector Ltd had no choice but to start proceedings against Malakoff for obtaining information about the machine by misrepresentation and fraud and it obtained an injunction to restrain Malakoff from exploiting the invention. Matthews then had to undo the damage done by Malakoff lodging the 'interference'. The cost of litigation was paid for by his wife and accounted to about $75,000. Malakoff claimed to have thought of the Sky Projector as far back as 1910 but could provide no proof of this claim. Because of the interference G.M. Sky Projector Ltd never really functioned and was unable to attract any funds for the construction of more Sky Projectors.

Matthews, desperate to save G.M. Sky Projector, entered into negotiations with various advertising companies including one such company called Messrs Fisher Foils Ltd of Hendon. But after a delay of three months where they 'dilly dallied', the company refused to have any further interest in the invention. Matthews then offered the directors, Colonel Williams and Mr Fisher, the British rights to the invention for £60,000, the value placed upon the Sky Projector, which they were prepared to accept, subject to a satisfactory demonstration. But he was unable to give such a demonstration because his demonstration apparatus was 'tied up by the Sperry Gyroscope Company for a debt of £25'.[14]

On September 23rd 1931 Matthews returned to the Felix Hotel, London, where he was lodging, only to be served with a bankruptcy order. He had been living on credit and hadn't paid his bills for months and had been dreading this. The petitioning creditors included Gioconda Biondi, the proprietress of the Felix Hotel, and Godfrey Philip Orde, a director of G.M. Sky Projector Ltd.[15] He owed Madam Biondi a debt for board and lodgings totaling £311. Matthews had used rooms and at the Felix Hotel as offices to conduct his business affairs. The cause of his insolvency was the failure of G.M. Sky Projector Ltd and the inability of the company to exploit the Sky Projector patent owing to lack of funds. There were twelve creditors in all, including several contractors who supplied him with goods and services, the Post Office, Count Valambrosia who had lent him £6,000, Mr. J.P. Evans who had advanced him a total of £1,386 and Major Mackay who lent him £1,737 between 1920 and 1923 for his work on talking films. His creditors met on October 19th 1931 to decide on the best course of action and Mr E. Burke was appointed the official receiver and acted as trustee of his estate. Matthews submitted his Statement of Affairs and accounts on December 23rd 1931 to the Bankruptcy Court, informing them that his debts amounted to £11,441 and declared that he had no assets with the exception of his 60% share in the Sky Projector.[16]

Of the £11,441, £4,285 was unsecured for cash advanced, hotel accounts, goods supplied, solicitors' costs and telephone charges.[17] Poor old Matthews, Christmas that year must have been a rather grim affair as court records show that he had to sell what little furniture he had along with his personal effects which raised the sum of £8.[18] So desperate were his finances, the court discovered, that he had borrowed £50 from Leslie Block, a money lender.

The date of his bankruptcy hearing was January 15th 1932 at 11 o'clock in the High Court of Justice, Bankruptcy Buildings, Carey Street, London.[19] Under cross-examination from Mr Bruce Park, Matthews confirmed that his name was, after changing it by deed poll in 1916, in fact Harry Grindell Grindell but he was known as Harry Grindell Matthews. Why he had changed his name remains a mystery. If he was trying to avoid his creditors, as a result of his previous financial undertakings, then surely he'd have made a more significant change to his name, not a repetition of his middle name.

He told the court that he had worked on various inventions for a number of years and that his work was financed by various interested parties to which, in return, he gave a stake in his patents. Matthews revealed the fact that between 1924 and 1930 he had no permanent address and stayed with friends, or at hotels, and that, with the exception of a few days in England he had lived in America, France and Germany; indeed throughout his life he lived at no less than eleven different addresses. Asked how he financed himself during this period he said that with the exception of the £3,000 he received as a consultant for Warner Brothers and £310 during 1930–31 as a director and consulting engineer to G.M. Sky Projector Ltd, he was dependent on his wife, Olive Waite, until September 1930 when she had divorced him in Reno, Nevada.[20]

When he told the court that he had returned from America with £100 in his pocket the court wanted to know what had happened to the £3,000 he had received for his consultancy work with Warner Brothers. Had he spent it on living expenses, could he have not lived more economically and paid some of his creditors? But the court realised that he was not guilty of extravagance since he had lived within his means and his standard of living was high whilst married to Olive Waite. He had valued his patent for the Sky Projector at £60,000, of which he had a 60% share which he thought money could be raised against and used to settle the unpaid bills with his creditors. His work on talking films had been partly financed by others as a joint venture and partly by some borrowed money and he claimed that his household and personal expenditure did not exceed £500 per annum.

Despite being declared bankrupt, on May 27th 1931, Matthews refused to give up and was convinced the Sky Projector had a future. Although a poor businessman himself he was adept at wooing potential investors, convincing them he had an invention worth a lot of money if only he could get financial backing. It wasn't long before he was able to get the financial backing he was looking for and was able to start up yet another small company called 'Luminastra'. It was set up in 1932 only to be, with depressing predictability, dissolved five years later in December 1938.[21] Unfortunately all records relating to Luminastra have been destroyed.[22] Details of the company's liquidation do appear in the *London Gazette*.[23] Just as with his previous companies, creditors met and decided that 'the Company cannot by reason of its liabilities, continue its business ... and accordingly ... the company [would] be wound up voluntarily'.[24] Having been previously declared a bankrupt Matthews wouldn't have been able to act as a director and so another director, B.H. Tucker, was appointed. With no records existing, the exact dealings of the company are a matter of speculation, but it is fairly certain the Luminastra was formed to acquire, finance and develop the improved Sky Projector. Matthews would on more than one occasion invent a device that showed considerable commercial potential, set up a company along with willing investors, and after varying degrees of success see his company go bankrupt. As with many inventors, he wasn't a particularly good businessman and had a poor grasp of financial matters, which is all too apparent in the vague answers he gave at his bankruptcy hearing and by the fact that he kept no financial records. His approach to all financial matters was lackadaisical to say the least. He would use money to pay his expenses and carry out research and development. He did have an expensive lifestyle, living in hotels, renting numerous properties, employing staff, draughtsman, solicitors and patent agents, and he travelled extensively. But before Luminastra went bankrupt he was able to build his mark two version of the Sky Projector.[25] Its powerful arc lamp was 25 feet in length and mounted on top a system of gears and sliders so that it could be swivelled easily in any direction. The whole apparatus was built on the back of a large truck; the whole thing looked like a giant telescope. This much bigger device was designed to project images onto clouds, and was made up of just one large tube and not two smaller ones set at right angles like the earlier 1926 version. Using a similar method to its predecessor, a beam of light illuminated a stencil but used two lenses, not one. The new device was enormous, with the beam from a huge searchlight in front of an elliptical mirror shining through the first lens, which then passed through the stencil set immediately after the lens and then through the second, focusing lens. The focal length of this, the mark two projector, was 15 feet. 'It swings in a circle by electrical machinery, searching for clouds on which to focus a picture many miles in extent, with simply a touch of a button. The projector is mounted on a gun-like platform on the chassis of the lorry, standing nearly twelve feet high, with a telescope-like barrel that can be raised almost vertical, or swung to point in any direction.'[26] The huge arc lamp would get terrifically hot and so fans were mounted at the side of the apparatus to blow a constant stream of cool air. The transparent slides with the image to be projected were mounted in round magazines that, via an electrical mechanism, slotted in front of the lens. Several plates could be employed to pass a variety of different images in turn in front of the lamp. This could be used to project a kaleidoscope of brilliant colours into the sky.

In the fading light of a wintry evening in 1932 excited shoppers hurrying across Hampstead Heath in London were amazed to see what looked like a searchlight sweep back and forth across the wintry sky. They looked on in surprise until suddenly the searchlight played upon a cloud and an image of an angel with outstretched wings appeared before their eyes. Frozen to the spot the bemused shoppers stood agog as the heavenly apparition hovered just under the snow-laden clouds only to fade away to be replaced by the words 'A Happy Christmas'. Cars stopped in the road and traffic chaos ensued. Matthews had set up his Sky Projector behind the high walls of Jack Straw's Castle, a well-known London pub on Hampstead Heath. Those enjoying a drink inside Jack Straw's Castle had noticed the activity in the back yard and wandered over to take a closer look. Realising that his cover was about to be blown Matthews slipped away in his car quickly followed by the lorry carrying the huge Sky Projector. This wasn't the only demonstration of the Sky Projector in England. Shortly after its debut on Hampstead Heath those enjoying an evening stroll in Blackheath were amazed to see two moons in the sky, only one of the moons appeared to have a winking, smiling face on it. The versatility of the Sky Projector was brilliantly demonstrated when a clock face, showing real time, was projected above the heads of those having a late evening stroll.

In the end financial success eluded the Sky Projector. Matthews had developed it as much as he could and, as with so many of his ventures he, had to shelve it. But he never gave up inventing. As is so often the case with pioneers he was simply ahead of his time, it was nothing more than that. Tesla, Baird and Marconi are all remembered as pioneering geniuses for their work: why isn't Matthews? But that wasn't the end of the Sky Projector – not quite anyway. Its potential as a method of advertising is obvious to all but the most shortsighted. When in 1937 Matthews and his biographer, Ernest Barwell, were in Germany on business, they were invited to meet with two senior members of the Nazi party, Goebbels and Goering. Goebbels, the minister for propaganda, thought the projector was ideal for Nazi rallies. But Matthews played down the capabilities of the projector saying 'messages are blurred and indistinct, while it is hopeless to project anything like a photograph'.[27] Later when back in England he was visited by a Spaniard claiming to be a member of the Falange political organisation who was interested in his Sky Projector. He told Matthews that General Franco, the Spanish dictator who overthrew the Spanish democratic republic during the Spanish Civil War, was interested in the Sky Projector for party propaganda purposes. But Matthews told him in no uncertain terms that he wanted no such involvement in such matters. That Spaniard had made a long journey to see the inventor for he was, at that time, living in a remote mountain laboratory deep in the Welsh mountains overlooking Swansea.

Notes

1. http://v3.espacenet.com/ patent number GB359125 (accessed 30/12/07).
2. http://v3.espacenet.com/ patent number GB254437 (accessed 30/12/07).
3. Barwell, p. 120.
4. Ibid.
5. www.history.navy.mil/danfs/s16/sperry.htm (accessed 23/08/07).
6. National Archives B9/1190.
7. http://v3.espacenet.com/ patent numbers GB272606, CA274165, FR620965 (accessed 30/12/07).
8. Ibid.
9. Barwell, p. 118.
10. Ibid., p. 120.
11. National Archives BT 226/4670.
12. National Archives B 9/1190.
13. Ibid.
14. Ibid.
15. National Archives BT 226/4670.
16. Ibid.
17. National Archives B 9/1190.
18. National Archives BT 226/4670.
19. National Archives B 9/1190.
20. National Archives BT 226/4670.
21. *London Gazette*, July 13th 1937.
22. Information supplied by Companies House.
23. *London Gazette*, June 29th, July 13th, December 3rd 1937 and November 18th 1938.
24. *London Gazette*, July 13th 1937.
25. http://v3.espacenet.com/ patent numbers GB408406, FR736708, US1862577 (accessed 30/12/07).
26. Barwell, p. 125.
27. Ibid., p. 127.

CHAPTER 8

The Mystery Man of the Mountains

O<small>NE BRIGHT SPRING</small> Morning in May 1934 a tall, smartly-dressed man knocked on the door of a small farmhouse on the Mynydd y Gwair Mountain in Craig Cefn Parc overlooking Swansea, Wales. The farmhouse was the home of the Jones family who ran a successful farming concern and the stranger at the door was Matthews, asking whether any lodgings were available. Glad of the extra income Mrs Jones offered him the use of a small back room and her daughter, Mabel, vividly remembers the time that Matthews stayed with the family during the spring and summer of 1934. She recalls that a chauffeur-driven car, which was a magnificent black Hudson Terraplane with the private numberplates GM 1, would come to collect him early in the morning, and drop him off later that same evening. He never ate with the family, preferring to have his meals in the 'Masons Arms' in neighbouring Clydach. Mabel remembers Matthews would often give her and her brothers and sisters half a crown each. 'He was a kind man and I particularly remember that he wore pure silk socks. His eyesight wasn't very good and he wore horn-rimmed spectacles with an eye patch.'[1] Before breakfast each morning her mother would take a jug of hot water into Matthews' room for his morning wash but on one occasion she discovered he wasn't there. 'My Mother found him having a wash at the cold tap outside. This was how my brothers and father used to wash and Mr Matthews wanted to be part of the family.'[2] Although solitary and independent by nature he found living with a family on a working farm rather to his liking, showing an interest in the farm, and he would often be found in deep conversation with Mabel's elder brother who was the intellectual one of the family. The Jones family wasn't immediately aware of what Matthews was doing in the area but it soon became known that he was, at that time, building a bungalow on the other side of the mountain and by lodging with the family he had a convenient base which meant little travelling about.

Owned by the Duke of Beaufort, Craig Cefn Parc was and still is a remote area of South Wales. The name means 'rocks behind the enclosure',[3] a reference to the rock material quarried from one of the numerous stone quarries in the area, it is a very isolated and windswept region. To the north, about one thousand feet above sea level, is a large area of common moorland. The region is mountainous, with the largest mountain being Mynydd y Gwair, meaning 'grass mountain'.[4] Clydach lies to the south and Ammanford to the north along with the remains of a Roman wall, and further still are the remains of an ancient castle.[5] Matthews chose this particular spot because it offered him 'an untouched background of space, secrecy and privacy'[6] and he would be able to continue his work undisturbed and 'devote himself to research, and pursue his investigations without curious eyes prying and breaking his solitude'.[7] He built his mountain bungalow, which he called Tor, meaning 'high', Clawdd, 'hedge or earth bank'[8] on top of the Mynydd y Gwair Mountain. Matthews chose this location not only for its rural isolation but also because he found that the remote wide open spaces soothed him and he would often indulge his passion for long solitary rambles over the moorlands. Matthews simply loved the remoteness of it all, the moorland pastures, wide open skies, the rolling mountainside and distant horizons. 'The only signs of man's influence on the area were the tall grey pylons stepping over the landscape in the distance, and the narrow rocky mountain road.'[9]

Both his arrival and the building of Tor Clawdd caused quite a stir. With many local labourers working there, including a well-known local character called Will Francis from nearby Felindre, rumours and wild stories of what was happening in the remote laboratory high up on the Mynydd y Gwair swept through the homes and pubs in the Clydach valley. Who was the stranger and what was he doing up there? On warm summer afternoons locals would make the long walk up the mountainside and, finding a good vantage point, would picnic hoping to catch a glimpse of the mysterious scientist who had chosen to live there in such isolated seclusion. What they saw when they got there was a large white bungalow being built, just in the shadow of the summit so as to avoid the worst of the weather, around which was a tall electrified fence with large steel gates. Local stories told of how he was working on a mysterious ray that could wipe out whole armies, inventing a machine that could control the weather, building a rocket that could travel to other planets and that he was also working on germs that caused a virulent disease. Wherever Matthews went he was incapable of living a quiet life and inevitably the local press became interested in 'the mystery man of the mountains'.

After about five months of building frenzy the bungalow was finished in the late autumn. 'He wanted it ready as quickly as possible,' Mabel Thomas recalled more than 70 years later. Designed and constructed to his own specifications, at a cost of £3,200[10] on two acres of land, Tor Clawdd made an incongruous sight on the Mynydd y Gwair Mountain with its smart white exterior contrasting sharply with the surroundings. The bungalow still stands today, its location in a small hollow making it difficult to see from the main road, and is a monument, indeed the only surviving monument, to Matthews' pioneering work.

Its outward appearance gave no hint of the cosy, well-appointed rooms within for inside there was a large, well-equipped laboratory, referred to as the long room, six bedrooms, two bathrooms, a lounge, dining room and a kitchen. Each room was well furnished, with the lounge having a leather three-piece suite, a 'Telefunken' radio, a piano and a large open log-burning fireplace above which was the Grindell Matthews coat of arms and motto *'Regardez les Cieux'* or 'Look to the Skies'.[11] The kitchen had an 'Echo' radio, a large American-style fridge and an electric cooking stove. With both mains electricity and its own power supply provided by a generator and transformer, and water drawn from a man-made well, it was the ideal place for Matthews to continue with his work in relative seclusion. Matthews was a qualified pilot and behind Tor Clawdd he had a small airstrip built so that he could land his private light aeroplane which was either a Foxmoth or Pussmoth. Mr Price was the contractor who worked with half a dozen men to build the air strip.[12]

When the house was complete Mabel would often go up to keep Matthews' housekeeper, Lilley Morris, company whenever he was away. She enjoyed staying with Lilley because 'the bungalow was well furnished, very comfortable and self contained with a lovely sitting room and I remember a lovely big cat. There were two bathrooms and hot water. Lilley was a wonderful housekeeper and a cook.'[13] Lilley wasn't the only person in the employ of Matthews for he also employed a handyman and gardener, Aubrey Morris, who kept a very tidy garden area, in the grounds of Tor Clawdd, to provide the household with fresh fruit and vegetables. Mabel and her sister used to walk across the fields following the overhead telephone wires, to take his milk up for him from their own farm. Surrounding the property was a tall electrified fence topped with barbed wire: Tor Clawdd guarded its secrets very well. 'At the gate there was an electric door bell and I think he had a camera installed later on.'[14] A lot of local people were wary and against him because of his association with the 'death ray' and Mabel remembers that some of the locals were surprised that they allowed him to lodge with them. They all knew him in the village as 'Death Ray Matthews'. But Mabel always remembered him as 'nice and very modest. He was a proper gentleman, very appreciative, and a man of great knowledge.'[15] Initially Matthews spent that first late autumn settling in, relaxing and getting to know some of the locals.

During 1939 and 1940 a young local lad, Ogwyn Norris, aged 12 at the time, used to go up to Tor Clawdd after school, and help Esmor Davies, who was then Matthews' assistant, push his aeroplane into a large shed behind Tor Clawdd. Ogwyn remembers Matthews as a 'nice man' who had a solitary air about him and never spoke Welsh. He recalls how Matthews was often accompanied by an unknown associate. After letting Esmor know when he planned to return Matthews and his associate would climb into the small, single wing, two-seater aircraft, with Matthews in the pilot's seat, taxi across the grassy airfield and head off into the skies towards Cardiff. Very often Matthews would fly off on a Friday evening to return on Sunday afternoon or Monday morning. At the time Matthews did still reside at a couple of addresses in London including 20 Bolton Street, W1 and Torrington Square in Holborn and could well have been heading for Croydon airport to keep an appointment in London. Ogwyn remembers being told by Matthews never to go into the laboratory which he clearly remembers as always being locked. Like other locals Ogwyn remembers stories of Matthews stopping engines of passing cars with an invention of his. 'When the war broke out the government came up and dug trenches to prevent aircraft from landing.'[16]

With his rare appearances in the local village, the coming and goings of his aircraft and rumours of mysterious figures seen going into his laboratory, it was the stuff of Biggles and the local children revelled in it all. One of those was local schoolboy Denzil Davies who would walk up the remote Mynydd y Gwair to try and catch a glimpse of the inventor. Audrey Watts remembers her uncle, a close friend of Matthews, being with him when he demonstrated a small device of his stopping a car travelling down the main road leading past his laboratory:

> My uncle remembers walking on the mountain and talking to Mr Matthews. He was sitting on the grass near the roadway with a sort of old fashioned 'radio' at his side. He told my uncle he was waiting for a car to pass by. They chatted for about three quarters of an hour and eventually a car, there weren't many on the roads in those days, came into view. Immediately Mr Matthews twiddled the knobs on the 'radio' and the car stopped. The bemused driver got out, lifted up the bonnet and fiddled with the car's engine, but the car remained silent. After a while Mr Matthews adjusted his 'radio' and the car started up again, which gave Mr Matthews great satisfaction. All he said was, 'Well that worked.' My uncle swears he saw this with his own eyes.[17]

A novel invention that Matthews was working on at the time was a dog whistle which was the result of his studies in to the effects of high-frequency sound. He was fascinated by the effect of watching high-frequency sound shatter glass and actually thought the principle could be used in warfare. According to Barwell, Matthews 'had some success with a sound beam and saw a cartridge fired by it'.[18] As well as the Jones family Matthews also got to know the Richards family very well and would call in of an evening for supper and would enjoy the family atmosphere. During those fireside evenings William Richards got to know Matthews very well. A few short years later William and his son-in-law, Doug Smith, would act as coffin bearers at his funeral. But William's daughter, Edna, fondly recalls her memories of Matthews when he used to call in at their farmhouse on the side of the mountain. She remembers her father telling her how he saw Matthews use a device to stop the engine of a car travelling down the road past Tor Clawdd and that he witnessed the arrival and departure of his aeroplane. A photograph from the time shows an aeroplane, reputed to be carrying VIPs, taking off from his private airstrip.[19] 'He was a very nice-looking man with lovely skin,' Edna recalls. 'He was a really nice gentleman and used to sit in my father's chair.'[20] She remembered his surprise when he turned up one evening to find the family toasting bread in front of the open coal fire and they would invite him to have some of their home-made stew, or 'cowell'. Edna was fortunate enough, on a few occasions, to go inside Tor Clawdd and like Mabel Thomas remembers how nice it was and how she saw 'some contraptions'. During one of her visits Edna was surprised to hear Matthews say that he had met with Winston Churchill in the past to discuss his work, only to be told that he was '100 years too advanced'.[21] After the war broke out Edna distinctly remembers a guard being posted outside Tor Clawdd.

There are recollections of Matthews being driven around the village in a large, black, chauffeur-driven car. He would appear, tall, smartly dressed in black with a patch over one eye, only to promptly disappear into the local grocer's shop. Emerging a short while later to find the local children had gathered excitedly around his black Hudson Terraplane car, he would hand out sweets to them.[22] The sound and sight of Matthews' aeroplane flying across the quiet skies of Swansea was quite a spectacle in the late 1930s and a young Roy Williams would often climb up to Tor Clawdd to catch a glimpse of Matthews landing his plane, an excursion that would often make him late for school where the cane was waiting for him! Roy also remembers an electrified fence and an army guard being placed outside Tor Clawdd.

One of the first of several distinguished visitors to call in at Tor Clawdd was the adventurer and explorer Colonel P.T. Etherton. Etherton was part of the 1933 Houston-Mount Expedition to fly over Everest. Financed by the millionaire Lady Houston and led by Air Commodore Douglas-Hamilton, a small formation of open cockpit biplanes flew over Everest: the highest altitude then achieved by an aeroplane. An accomplished pilot and explorer, Etherton had travelled in Siberia and flown over both the Himalayas and the Andes. Entertaining his close friend, Matthews discussed with Etherton the future possibility of another world war and the current state of the nation's defences. They were both of the same opinion: that Britain would be vulnerable to invasion unless she could effectively defend herself from not only airborne attack but also submarine attack. In an article written for the *Sunday Express* telling of his visit to Tor Clawdd,[23] Etherton was to say that Matthews was continuing his work with submarine detection, but also looking into new lines of research which included a ray to kill germs, an aerial defence scheme for London and a rocket aeroplane. He also observed that Matthews had 'machinery that will transmit rays strong enough to kill a rat sixty feet away or stop a motor car'[24] – the device which had been seen in action by Edna Bodycombe's father and Audrey Watts' uncle. 'He is a genius, probing the secrets of electricity.'[25] But with the mystery man on the mountain, living in seclusion with the comings and goings of his light aeroplane and rumours of fantastic experiments and inventions, the press inevitably began to take an interest and it wasn't long before a steady stream of inquisitive journalists would climb the mountain to Tor Clawdd. Having made the long and bleak journey up the mountainside they would be confronted by an electric fence and a locked gate, unable to gain entry. Matthews' growing notoriety also brought several people who could only be described as eccentric, to his door with their own, rather novel, inventions. Some of them did however impress the inventor, including a 'sticky bomb' and an anti-tank gun.

In 1934, the same year that Matthews built Tor Clawdd, some of the finest scientific minds in the country gathered in London to meet at the Committee for the Scientific Survey of Air Defence (CSSAD). Chaired by the distinguished scientist Sir Henry

Tizard, scientific advisor to the government, the CSSAD was to examine the threat posed by the German Luftwaffe and how best to defend the shores of the United Kingdom. Sir Henry and his team looked at a variety of methods including barrage balloons and acoustic or sound mirrors. Sound mirrors were used during the Second World War to detect approaching enemy aircraft in the days before radar. Built from concrete and shaped in such a way as to focus and amplify sound waves, not only could they detect the range of enemy aircraft to a maximum distance of 20 miles, but also their direction. Unfortunately they were, due to their restricted range, of limited use. But after much debate they seriously considered, not for the first time, the idea of the 'death ray' and the government offered, with an uncanny echo of Matthews' experience back in 1924, £1,000 to any person *'who could build a death ray that could kill a sheep at 100 yards'* (author's italics.)[26] However nobody came forward with a practical device and the prize went unclaimed. Throughout the 1930s the British Government were, just like they were in the 20s, receiving intelligence reports that suggested Nazi Germany had made a 'death ray' which used radio waves that could destroy aircraft. The CSSAD looked into the claims. Major H.E. Wimperis from the CSSAD and director of scientific research for the Air Ministry, visited a young Scottish electrical engineer, Robert Watson-Watt, working at the National Physics Laboratory in Teddington and wanted to know if the 'death ray' was a practical proposition that could be used against enemy aircraft as Matthews had proposed over a decade earlier. Wimperis had witnessed Matthews' electric beam back in May 1924. Sir Henry Tizard said of Matthews, 'Personally I should say that the man is trying to bamboozle you, but don't let that affect you at all!'[27] Robert, later Sir, Watson-Watt scrutinised the idea of creating a beam of electromagnetic energy to destroy aircraft. Working with his assistant, Arnold Wilkins, they discovered that such a 'death ray' was not possible with the available technology and so the idea of developing one, along with the possibility that the Germans had such a device, was dropped. But his work had led him to look more closely at the phenomenon whereby aircraft flying past radio transmitting aerials belonging to the BBC caused a disturbance in the radio signal being transmitted.

Receiving aerials were picking up an incomplete signal. Realising that the radio waves were bouncing off the passing aircraft Wilkins and Watson-Watt appreciated that radio waves could be used to detect the presence of aircraft and so Radio Detection and Ranging or radar was born. Watson-Watt and his assistants did an experiment on February 26th 1935 at Daventry in Northamptonshire which proved that it was indeed possible to use radio waves to detect and track moving aircraft. Another member of that team lead by Sir Robert was Edward Appleton who used radar to prove the existence of the ionosphere, the uppermost part of the earth's atmosphere, that enables radio waves to be transmitted around the world by a phenomenon called 'wave propagation'. Appleton was a close friend of Baird and went on to win the Nobel Prize for his research. Radar would play a pivotal role in the Battle of Britain, enabling the Royal Air Force to scramble fighter aircraft so effectively that those referred to by Winston Churchill as 'The Few' were able to defeat the German Luftwaffe in the face of incredible odds in the skies over southern England. One of Matthews' strengths was recruiting help with his work. One such willing helper was a keen amateur electrician called George Howell who worked briefly with Matthews on his electric beam and later went on to join Royal Air Force where he worked on radar. After the war he ran a successful small business repairing TVs and radios in Swansea.[28]

Although no documentary evidence exists it is quite possible that Matthews could have been in contact with the CSSAD throughout the 1930s for it has been rumoured that government officials did make regular visits to his mountain laboratory. There was certainly a regular exchange of correspondence between Matthews and the government during this period. Widely considered to be impractical by the establishment at that time, Matthews' 'death ray' prompted questions, the answers to which brought about radar, and modern radar devices are currently being developed as weapons. For some years it has been known that radar systems are capable of generating enough energy to burn the electronic valves of enemy aircraft. Modern radars called Active Electronically Scanned Array (AESA) radars produce bursts of energy and can be used as jamming devices against enemy communication systems. Presently modern military services are now looking at using AESA technology to attack an enemy's radar systems, burning the radar's antenna. Scientists think that the energy can also be focused onto a range of targets, such as a missile.[29] But 'death ray' technology doesn't end there. An American defence company, Raytheon Co., has spent millions of dollars developing a 'heat beam' for the US military. This device focuses a small 'wave energy beam to induce an intolerable heating sensation'.[30] The weapon is designed to be non-lethal and to assist armed forces with peacekeeping duties such as crowd control. Raytheon Co. is planning a smaller version to be used by civil authorities. Current work and developments in this field show that Matthews had a prescience which until now has gone unacknowledged. Matthews was indeed visionary.[31]

On dark, wintry nights, Matthews would use his Sky Projector apparatus to help his assistant Esmor down the mountainside: 'He used to put the searchlight on, it was pitch black, and all you could see from the village was this man walking in the light. He kept the searchlight on until Esmor got from Matthews to the small road. Once he got there he used to switch the light off.'[32] Later, after Matthews' death, Esmor himself was to die a relatively young man, in his late forties, of cancer. Matthews would on occasions be seen around the local villages of Clydach and Ammanford. He enjoyed calling in to the Masons Arms, the village pub in nearby Rhydypandy, where he became very friendly with the landlord, Billy Bass, his wife Rose and their three children Bette, Dillwyn and Gyn. Local legend has it that he secretly met Churchill there to discuss his defensive weapons. Matthews would on numerous occasions play darts and enjoy a pint of ale with the locals. Various documents and letters held in the Swansea County Archives[33] offer a fascinating and detailed insight into Matthews' character and personality. Whenever he was seen by locals he would always be immaculate in appearance, wearing a black suit with a silk shirt, tie or cravat along with a velvet Stetson hat and hand-made shoes. He was a tall man, 6'4" in height, with broad shoulders, a high forehead and blue eyes. He loved nature and the outdoors and particularly enjoyed rambles over the nearby mountains when he could be seen striding out wearing plus-fours. In the laboratory he was meticulous in his habits, always wearing leather gloves when handling electrical equipment, and he kept meticulous notes. His laboratory was always kept incredibly 'tidy to the point of being clinical and packed with electronic equipment of every description'.[34] When working on a particular problem he required very little sleep and would often work through the night. He was a very welcoming host to visitors at Tor Clawdd, offering them good meals or a large Sunday lunch, a particular favourite of his, prepared by his housekeeper, Lilley Morris. Sitting with his guest in front of a log fire they would relax and listen to plays and music on his Telefunken radio, have a game of cards or darts or play the piano. He was an inveterate tea drinker and a chain smoker. Matthews found Lilley to be an excellent housekeeper, keeping his domestic affairs in order whilst making Tor Clawwd a cosy and comfortable home for him. He loved animals and had two cats, Toots and Pudge, of his own. Those who knew him closely described him as having a magnetic personality with very tidy habits. Although when relaxed he could be very jovial, he could never tolerate fools and could be very blunt and impatient. He enjoyed visiting the Jones and Williams families and although constantly preoccupied with his work he would rarely discuss it and never spoke about his own family.

In January 1938 Matthews married for a third time. The object of his affections was Madam Ganna Walska, and they were married on January 22nd 1938, in what was a very low-key affair, at the Holborn Register Office, central London. Their marriage certificate records both of them as 'being of independent means', with Matthews' occupation as 'Electrical Engineer'. Eleven years her senior, Matthews was Ganna's fifth husband and immediately after the wedding ceremony they honeymooned together in St Moritz, Switzerland. Born Hanna Puacz in 1887 in Poland, Madam Ganna Walska was a well-publicised socialite and opera singer. A larger-than-life personality, she wrote her autobiography *Always Room at the Top*, in 1943.[35] 'Dedicated to all those who are seeking their place in the sun', her autobiography is a self-indulgent, ponderous, yet articulate book, but does offer an interesting glimpse into her brief marriage and relationship with Matthews. Ganna studied music from an early age and became an opera singer, with somewhat limited success. Struggling to be taken seriously, reports appeared in the press ridiculing her ability as an operatic singer. Travelling extensively throughout Europe and America during the 1920s, she attracted several admirers amongst the rich and famous including Charlie Chaplin, Richard Strauss and Franklin D. Roosevelt. Not averse to getting married, which she did six times, she amassed a personal fortune that led her being called the $25,000,000 bride. With her wealth she developed a penchant for collecting fine objets d'art. One of her more wealthy husbands was Harold McCormick, who owned a large agricultural company, and bought her the Champs Elysées theatre in Paris.

In 1941, with the help of Theo Bernard, her sixth and final husband, Ganna used her wealth to buy a large estate in Santa Barbara, California and set up 'Lotusland', one of the world's largest botanical gardens with an impressive collection on rare and exotic plants.[36] She died, aged 97, on March 2nd 1984, leaving her garden along with her personal fortune to the Ganna Walska Lotusland Foundation.

Matthews had first met Ganna Walska at the Authors' Club in London. A close girlfriend of Ganna, Gita, thought they would make a good match and invited Matthews to Covent Garden, where she and Ganna were going to see an opera by Debussy, *Palléas and Mélisande*. A fluent speaker of French, Matthews was smitten and shortly after declared his love for Ganna, and proposed to her. But initially the feeling wasn't mutual with Ganna being somewhat reluctant. Unhappy at her procrastination, his woes were added to when he heard his sister was dying of cancer. Ganna said 'he told me that he had been working on an invention to conquer the terrible malady.'[37] Matthews became increasingly depressed at his sister's terminal condition and sleepless nights became a problem. Over the next few months his own health seem to deteriorate and he suffered from perpetually chronic headaches, not helped by the fact the he had to continually wear an eyepatch. He also noticed his hands trembled, particularly when he was tired, and along with his poor eyesight he found it difficult to shave and groom properly which resulted in his appearance becoming increasingly dishevelled, with Ganna noting that suddenly he 'looked older than his age'.[38]

Matthews had for some while suspected he had a heart condition, for on top of everything else he found his hands and feet grew cold quickly and he often felt pins and needles in his limbs. He was urged by Ganna to visit the doctor, who confirmed he had high blood pressure and prescribed tablets to treat his condition. All this affected his ability to work effectively which made him frustrated and restless, and he would become argumentative with close friends – even his biographer, Ernest Barwell. His relationship with Ernest became particularly strained and it was only because he was under contract to complete the biography that the job was completed at all. Ruminating on his past, Matthews realised how he had given everything: his health, his money, indeed his life and all for what? A series of rejections and a reputation as a crank and fraudster. He became very negative in his outlook and 'considered himself a victim of jealousy, injustice, and treachery'.[39] He would launch into verbal diatribes about his treatment at the hands of the authorities and how 'the government was not willing to collaborate with him'.[40] All this and the loss of a sister resulted in the inevitable and he suffered a nervous breakdown.

But despite everything, Ganna found herself falling in love with him. When he was himself he was confident, self-assured and intelligent, with a wonderful sense of humour, and she eventually agreed to marry him. Ganna was not only a very supportive wife but also a very wealthy and willing patron of his work. She paid for the construction of a purpose-built laboratory at her home, the Chateau De Galluis, just outside Paris. Located on a vast estate with extensive grounds, Matthews convalesced for weeks at a stretch, particularly during the winter months, and continued his work without financial worries. As his health gradually returned he divided his time between Tor Clawdd, London and Paris. Ganna moved in some elevated circles, mixing with the famous, the wealthy and the influential, and through his wife Matthews came into contact with foreign diplomats on the Continent. What is very interesting is that Ganna records that 'in London he was not allowed to take two steps without being escorted for the sake of protection by a Major in the Intelligence

Service' and that 'he could not visit [her] in France as his government was afraid he would be killed'.[41]

However, shortly after they were married problems in their relationship began to emerge. Matthews was forever preoccupied with his work and would again become quite obsessive. He did have a jealous streak where Ganna was concerned; she was wealthy, vivacious and considered to be quite a beauty. She mentions in her autobiography how he would on occasions enter into uncontrollable rages, always suing someone for calumny (misrepresentation) and constantly defended his good name through his solicitors. He had a love of publicity, enjoyed reading about himself in the newspapers and 'he himself gave out the material for them'.[42] The marriage was destined not to last very long and shortly before Matthews died they parted company. Edna Bodycombe remembers how her sister recalled the last time anyone saw Ganna at Tor Clawdd: 'One morning a taxi went up to the Richards Farm, as there was no serviceable road up to Tor Clawdd and "the Countess" came down to the taxi, jumped in and that was the last we saw of her.'[43]

Notes

1. Conversation with Mabel Thomas, née Jones, 15/03/07.
2. Ibid.
3. 'A Castle on the Common', unpublished leaflet by I. Richard, 2003.
4. Ibid.
5. Ibid.
6. Barwell, p. 11.
7. Ibid., p. 127.
8. Conversation with Ioan Richard, 31/09/06.
9. Unpublished manuscript, *The Visitors to the Mountains*, Swansea County Archive D/DZ 346/16.
10. 'Statement of Assets', Swansea County Archive D/DZ 346/14.
11. Barwell, p. 130.
12. Conversation with Roy and Myra Williams, 15/03/07.
13. Conversation with Mabel Thomas, 15/03/07.
14. Ibid.
15. Ibid.
16. Conversation with Ogwyn Norris, 15/03/07.

17. Letter from Audrey Watts to Arthur Barwell, Swansea Archive D/DZ 346.
18. Barwell, p. 134.
19. *Clydach Photographic Souvenir of Old Clydach Vol. III* (Clydach Historical Society, 1996).
20. Conversation with Ogwyn Norris, 15/03/07.
21. Conversation with Edna Bodycombe, née Richards, 15/01/07.
22. Conversation with Roy and Myra Williams, 15/03/07.
23. Barwell, p. 130.
24. Ibid., p. 131.
25. Ibid.
26. 'Radar and the Death Ray', *Fortean Times*, October 2003, available at: www.forteantimes.com/features/articles/195/radar_and_the_death_ray.html (accessed 28/12/07).
27. Confidential letter from Tizard to Wimperis, National Archives File Air 5/179.
28. Letter held in the Swansea County Archive, D/DZ 346/10.
29. Fulghum, David A. and Barrie, Douglas, 'Directed Energy for Missile Defence – Radar Becomes a Weapon', *Aviation Week*, September 4th 2005, available at: www.aviationweek.com (accessed 21/08/07).
30. http://boston.bizjournals.com/boston/stories/2004/11/29/daily30.html (accessed 21/08/07).
31. Ibid.
32. Conversation with Jeff Morgan, 15/07/07.
33. The Grindell Matthews Collection, Swansea County Archive D/DZ 346/9, D/DZ 346/10, D/DZ 346/11, D/DZ 346/14.
34. Swansea County Archive D/DZ 346/14.
35. Walska, Ganna, *Always Room at the Top* (Richard R. Smith, 1943).
36. www.lotusland.org/history.html
37. Walska, p. 438.
38. Ibid., p. 440.
39. Ibid., p. 441.
40. Ibid.
41. Ibid., p. 442.
42. Ibid.
43. Conversation with Edna Bodycombe and Jeff Morgan, 15/01/07.

CHAPTER 9

Breakfast in London, Lunch in New York

LATE ONE AFTERNOON in the autumn of 1935 three figures on the Mynydd y Gwair Mountain, just behind Tor Clawdd, were struggling to get a small, home-built aircraft, known as a 'Flying Flea' airborne. Making one last attempt, as the light was beginning to fade and a cross wind was gathering, the pilot, Gwillym Price desperately clutched the controls. With the small engine whining under the strain of the throttle, the aircraft began to creep forward as the two other figures, breathing heavily with the strain, pushed. Then as the pilot opened the throttle still further and pulled back on the main control lever, the engine whine increased and to a chorus of cheers the 'Flying Flea' was airborne. The two men pushing the 'Flying Flea' were Matthews and Jim Hill, whose garden shed the aircraft was built in. Matthews had become friendly with both Gwillym and Jim who were both electrical engineers from the village. Gwillym worked for the local council and had visited Tor Clawdd to check and certify the electrical installations that Matthews had there and they discovered a mutual interest in electronics and aeronautics. He explained to Matthews that he was, along with a friend, trying to build a 'Flying Flea' and they were having some difficulty with its construction. Matthews' offer of help was eagerly accepted and the three of them set about completing the project. Gwillym's daughter, Helen MacDuff, recalls that her father 'had a high regard for Matthews and remembered his disappointment at the government's rejection of the "sting ray". He used to encourage my father to bring the "Flying Flea" up to Mynydd y Gwair for trials.'[1] Later both Gwillym and Jim went on to work for the Westland Helicopter Company at Yeovil. Details of how to build a 'Flying Flea' – the name given to a range of home-built aircraft designed by a Frenchman called Henri Mignet – were available in the monthly magazine *Practical Mechanics*.[2] Using his experience as a furniture manufacturer, Mignet's 'Flying Flea' was made from nothing more than wood and fabric with the whole thing held together with nails and glue, with a wingspan between four and six metres depending on construction and powered by a small two-cylinder engine of the constructor's choice.[3] Its 'construction does not call for a greater degree of skill nor of tools or equipment than is possessed by most amateurs'.[4] Mignet wrote a book in 1935 – *The Flying Flea – How to Build and Fly It* – giving all the necessary instructions and charts required to build and fly his aircraft. The local press reported details of the aeronautical success of Matthews and his friends.[5] Matthews' mountain laboratory Tor Clawdd provided the perfect place for the flight tests of their fragile, home-built aircraft where, according to the press, 'flights were undertaken near Mathew's [sic] small Radar Research unit'.[6] 'Radar Research unit' reflects the myths, which would only deepen with such details appearing in the local press, which surrounded Matthews and what he was reportedly up to at Tor Clawdd.

Matthews also became friendly with the local bobby, P.C. Jack John, who used to make regular visits to Tor Clawdd to have a chat with Matthews. Their friendship was so close that Jack would act as coffin bearer at Matthews' funeral. It is not certain if their friendship was purely social or if John had an unofficial role overlooking Matthews' security. 'A German government official did make a pre-war visit when they were interested in his "death ray" and projector. They wanted to see what he'd got and Goebbels did have an interest in his Sky Projector.'[7] It is known that the government was keeping its eye on him and during the late 1930s with international tension building and the rise of fascism in Europe, and with their dim view of Matthews, thinking he was a racketeer prepared to sell to the highest bidder, they just wanted to keep tabs on him. After all, he had worked for foreign companies and was well known on the Continent in the columns of the newspapers. Matthews asked John if he would help with the digging of a small rectangular 'pool' in the grounds of Tor Clawdd. Matthews claimed this was a swimming pool but witnesses claimed it was too small for such a purpose and was more than likely used as a testing area for his work on underwater detection apparatus for submarines. Documents in the National Archives[8] clearly show that during the 1930s Matthews was still working on submarine detection, an idea he first offered the British Government in 1917. Even though the Admiralty had written to Matthews to inform him that 'no further benefit is likely to result either by continuing these experiments or by prolonging the correspondence on the subject'[9] he returned to developing his submarine detection apparatus. Although the authorities weren't keen on his idea he never abandoned it and had written a letter to the Admiralty before moving to Tor Clawdd, in February 1931, informing them that, after making improvements to his apparatus, he could detect the presence of a submerged submarine 'at a range of over twelve miles'[10] and improved the apparatus so much that he was confident enough to claim that he could not only track the movement of a submarine but also pinpoint it to within 100 yards. 'My method will enable an officer on the bridge not only to hear but also to see on a small screen the presence of a submarine and visibly to witness its direction and the closing of the range.'[11] He asked if he could have the use of a submarine so that he could 'substantiate my claim to the satisfaction of your experts'.[12] But, just as before, he couldn't interest the Admiralty or get them to take a closer look at his invention and he'd long suspected, as was the case, that the Admiralty had wanted nothing more to do with him. Confidential minutes in Admiralty file ADM 116/4766 record that it is 'undesirable for the Admiralty to have future dealings with Mr. Matthews'. However the Director of Scientific Research asked the War Office whether or not Matthews was on a 'list of persons with whom no correspondence is to be held' and if there would be any objection to inviting him to put details forward regarding his apparatus for detecting submarines. The War Office replied that there was 'no great objection to inviting this gentleman to submit his new invention in the usual routine manner. Very careful consideration would need to be given before entering into definite negotiations with Grindell Matthews or allowing him to carry out experiments in H.M. Ships.'[13] His attitude to the submarine detector must surely prove him as a genuine and a committed inventor. After submitting details of his improved apparatus the Admiralty replied, on March 10th 1931, to say that they did 'not desire to investigate your system of submarine detection'.[14] He wouldn't let it drop. Convinced that another conflict with Germany was looming and anxious about Britain's defences, or rather lack of them, he continued with it under his own steam. Placing his apparatus into his car he travelled down to the Mumbles area of Swansea where there were submarines in the Bristol Channel. He worked with a 'brilliant young radio engineer, George Howell'. Howell was later to write from the Middle East, 'I am convinced that G.M.'s detector is the only thing that can beat the U-boat. I am sure that by now the Admiralty has taken it up, and thus the threat to our merchant shipping will have been wiped out.'[15] Apparently the authorities were aware of his activities and sent an officer to enquire what he was up to, and he was 'amazed at the ease with which the submarines were picked up and their courses charted'.[16] But he was unable to interest the upper echelons of the Admiralty.

Knowing that the Admiralty wanted nothing more to do with him, Matthews took a different approach and his next move was to act through a third party. He employed the services of a firm of solicitors: Messrs. Sheard, Breach and Co., who wrote to the Admiralty, on Matthews' behalf, asking whether it would be possible to use an abandoned coastguard hut near Stokes Bay, where there was a torpedo range, to house submarine detection apparatus for a trial. Matthews could use his apparatus on the submarines he knew to be stationed there. This was to be the start of a remarkable exchange of letters between Matthews and the authorities. Not knowing that the solicitors were acting on behalf of Matthews, the Admiralty referred the matter to the town clerk, Mr H.R. Mangnall of Gosport, who was responsible for the coastguard hut. The town clerk had no objections, and asked the Admiral of nearby Fort Blockhouse to let the solicitors know where the keys to the hut could be obtained. However the Admiral was not aware of any official research being carried out in the area, made further enquiries, and asked the town clerk not to hand over the keys to the hut in the meantime. Intrigued by the request, the Admiralty wanted to know the identity of the person requesting the use of the hut and the solicitors acting for Matthews wrote to the Admiralty stating that 'our client is a British subject by birth ... We can personally speak as to the bona fides and we have not the least hesitation in stating that we are not aware of any reason why our Client's request should not be granted ... It is our Client's intention to communicate with the Admiralty immediately the apparatus is ready for a test.'[17] The documents held in the National Archives relating to this show that the hut didn't actually belong to the Admiralty and therefore had 'no real locus standi. It does however afford a convenient opportunity of ascertaining further details of the purpose for which the hut is needed'.[18] The solicitors were asked to 'disclose, in confidence, the full name of their client, in order that he may be looked up' and a government representative was sent to interview the solicitors. He confirmed what the Admiralty had suspected, that the request had come from a Mr Grindell Matthews, and so as far as they were concerned the matter was at an end. They wrote to the solicitors refusing permission for their client to use the coastguard's hut. Writing back to the Admiralty, the solicitors informed them that

> Our client has handed us certain correspondence which passed between him and the Admiralty in the year 1917 and from a careful perusal of his dossier it would seem that the refusal of the required permission relates back to the enquiry held at Victory House in that year with regard to certain experiments at Barry.
>
> If this is so, we feel that our Client has been treated unfairly and having regard to the time and money our Client has spent on his discovery it is not possible for him to leave matters as they are. Although his loyalty will not be exhausted by misunderstandings it is not his intention to leave matters in this unsatisfactory state.[19]

Discovering that Matthews was Messrs Sheard, Breach & Co.'s client, the request was refused by the Admiralty in March 1934 in a letter marked 'Confidential.'[20] So as was his experience in 1917, Matthews simply couldn't get the official support his work merited. The Admiralty was no longer interested in his system of submarine detection. But Matthews wasn't that easily disheartened and his tenacity surely must prove his determination and faith in his submarine detection apparatus. A crank would have dropped the idea in the face of such rejection but a genuine inventor keeps going in the face of such circumstances. In March 1937, Colonel Frank Hilder got involved and tried to support Matthews in his return to submarine detection. Captain W.E. Parry at the department of Scientific Research & Experiment said that 'Grindell Matthews is a fraud.' But unconvinced, Col. Hilder, an MP and businessman, who had heard of what Matthews was trying to do, was frustrated by the Admiralty's' failure at taking up new ideas. Parry did put Hilder into contact with the Director of Scientific Research (DSR), because 'I felt that it just might be possible G.M. had some idea, which perhaps should be known'[21] but Hilder failed to interest the government. The DSR said that 'it is highly improbable that Mr. Grindell Matthews has anything of value to communicate'.[22]

Lady Lucy Houston, one of the richest women in England during the 1930s, had heard of Matthews, through their mutual friend, Colonel Etherton, and his work with wartime inventions, and arranged a meeting with him at her home on Hampstead Heath, London. A philanthropist, eccentric and adventuress she was, prior to the First World War, a suffragette. Married three times, her third husband, the shipping magnate Sir Robert Houston, died in mysterious circumstances in April 1926 leaving her an estimated £5 million in his will. Shortly after her husband's death she became paranoid and her behaviour became slightly erratic. According to the *Oxford Dictionary of National Biography* she had, during 1927, 'secret, flirtatious meetings at the Treasury with Winston Churchill'.[23] In 1931, after the Air Ministry refused funding, she financed, to the tune of £100,000, the British entry into the Schneider Trophy Competition to build and race seaplanes. Supermarine, the British aircraft manufacturer, designed and built the world's fastest seaplane, the Supermarine S.6B, winning the Schneider competition for the third consecutive year and thereby the trophy. The design team was headed by Reginald Mitchell, the aeronautical engineer, whose genius lay behind the legendary Spitfire. It was Lady Houston's generous financing that, in the face of government apathy, enabled the British aviation industry to keep pace with the developments happening on the Continent. Reportedly the government refused an offer of £200,000 from Lady Houston to finance the army and navy.

When they first met, Lady Houston was impressed by Matthews who struck her as a maverick: she saw in him a kindred spirit and they discussed his method of detecting submarines. Alarmed but not surprised to hear how the government had treated him, she offered him her wholehearted support. She couldn't understand why inventors like him received such scant support from government departments. Lady Houston suggested that, in light of the government's refusal to assist him with his trials for submarine detection, he use her private yacht *Liberty* to sail to the French naval base, Cherbourg, and continue the trials that he had abandoned in 1917. She proposed that if they moored *Liberty* close to the naval base Matthews could use his apparatus to detect the movements of submarines in the area, and arrangements were made for *Liberty* to sail to Swansea to collect Matthews and his equipment.

En route to Cherbourg, Lady Houston fell ill and the yacht had to be moored in Shell Bay, at the entrance to Poole harbour, where Matthews and Lynes took the chance to set up their detecting apparatus and were able to detect the propellers of passing ships up to 'three-quarters of a mile away' and 'voices and messages on various land telephones'.[24] Still feeling unwell and not wanting Matthews to proceed to Cherbourg without her, Lady Houston request they all remain at Shell Bay for a little while longer. With a shared sense of humour, Matthews spent the time regaling Lady Houston with stories of his work and in particular the one and only night he spent at her Majesty's pleasure in Brixton Prison. It appeared that a man acting as a secretary and manager for Matthews had, unknown to the inventor, withdrawn £800 from a business account. Discovering the withdrawal, Matthews promptly dismissed him from his employ but the disgruntled secretary demanded the return of £1,000 he had paid to Matthews for 1,000 £1 shares. Matthews refused and the ex-employee took the matter to court. In court Matthews claimed that the ex-employee had valid ownership of shares worth £1,000 but the share certificate had, said the judge, 'not been stamped'[25] and he therefore did not accept it as evidence.

But it was the plaintiff's duty to have done this when he was in Matthews' employ. The judge didn't accept this despite the plaintiff admitting that it was his signature on the transfer of shares certificate and consequently Matthews was arrested for the non-payment of the £1,000. By the time the judge had passed the verdict it was too late for Matthews to lay his hands on that amount of cash and so he spent the night in Brixton prison. Barwell wrote of his stay that Matthews 'was treated not as a prisoner but as a visitor and the warders waived regulations'[26] whereby he was allowed to smoke and wasn't made to have the compulsory bath upon admission. They had read about Matthews in the newspapers and felt they had a 'celebrity' in their midst. The sum of £1,000 was immediately paid the following morning and the matter was settled.

Meanwhile Colonel Etherton, a close friend of Lady Houston's since 1933 when she funded his flight over Everest, paid a visit to *Liberty* to see how things were progressing – only to find that she was still rather ill and Matthews was unable to continue with his experiments. Lady Houston was very apologetic about causing the delay which he must have seen as very irksome, but told Matthews that when she felt better he could have the yacht to himself. Although extremely grateful for this gesture Matthews simply couldn't waste any more time and with no prospect of her recovering sufficiently within the next few days, he decided to pack up his equipment and return to Swansea.

Matthews wasn't the only inventor working on submarine detection during the interwar years. In August 1936 Mr Eric Vernall wrote to the Admiralty with details of his 'Dyne' submarine detection apparatus. Claiming his apparatus was electrical, could detect a submarine at a distance of five miles and was in no way related to hydrophone devices, he requested facilities and officials to witness a demonstration. The Admiralty were prepared to examine his invention and provide facilities for a demonstration if, and only if, he was prepared to disclose particulars of the invention with an explanation on how it worked. He met with officials later that same month to discuss his work, but not revealing any specific technical details, the Admiralty were suspicious and ordered inquiries to be made in to the particulars of this gentleman. The subsequent report into Mr Vernall stated that it 'would be unwise to enter into any negotiations with him on the basis he proposes. There is a distinct danger that publicity of an undesirable kind would be given to any dealings with the Admiralty and is also likely that Vernall would endeavour to use the Admiralty as a means for repairing his financial situation.'[27] Admiralty files record that Matthews 'who is on the Admiralty black list'[28] was a colleague of Mr Vernall, and had 'given faked demonstrations to government representatives and used the occasions for advertising purposes and, to avoid a repetition of this, no further negotiations should take place with him',[29] and so the matter was dropped.

So Matthews wasn't the only one on the receiving end of such treatment. The Admiralty were right to be cautious when dealing with inventors but what had they to lose by sanctioning an official demonstration? The government certainly had a close eye on Vernall, even to the extent of spying on him, for there was a detailed report compiled about this man and his activities.[30] The report contained details of his business activities and his comings and goings at his offices in Regent Street, London. The Admiralty were under the impression that he had provided false references about his business activities in the past and owed rent on a previous office he had used. What may have particularly alarmed the government was that he had recently renewed his passport for all European countries. Could the government have thought that he might have been a spy?

Mr Vernall knew of Matthews and his work and contacted his assistant, Lynes, at his London-based manufacturing firm. Lynes replied stating that indeed Matthews and himself had been working on submarine detection and 'witnessed such positive results that the underlying principle is, in my opinion, now proved sound'.[31] Perhaps Mr Vernall was hoping to collaborate with Matthews and Lynes to bring the submarine detection apparatus to fruition but since the government wanted nothing to do with himself or Matthews the matter simply stalled. During the 1930s Lynes was still working with Matthews and it's interesting to note that 'B.J. Lynes, Ltd', a small experimental and engineering manufacturer, listed their interests in 'apparatus for talking films'.[32]

Between the beginning of the twentieth century and the

outbreak of the Second World War, the development of the aircraft had been swift and remarkable. The Wright Brothers achieved the first ever powered flight in December 1903 in North Carolina. Then just two years later they were able to fly continuously over a distance of 39km. July 1919 saw the Frenchman Louis Blériot fly across the English Channel from Calais to Dover in just 37 minutes. Biplanes such as the Sopwith Camel, with fixed machine guns, which were widely used for warfare and reconnaissance, appeared during the First World War. The interwar years would witness an explosion in aircraft technology with planes no longer being built from wood and fabric but aluminium with hydraulics, and engines becoming far more powerful and reliable. These developments allowed aircraft to travel much further – so much so that on June 14th 1919 John Alcock and Arthur Brown flew non-stop across the Atlantic from Canada to Ireland. Then the 1930s saw the dawn of the jet age when Frank Whittle, who was later knighted for his work in aeronautics, patented a design for a turbo jet that later powered the Gloster Meteor, the first British jet fighter. But the Gloster Meteor wasn't the world's first jet engine however: that distinction goes to the Heinkel He 178 that first flew in August 1939. So 36 years after the Wright Brothers made that first heavier-than-air flight, jet engines were able to achieve speeds in excess of 400mph.

Matthews realised that these developments in aeronautics would change the face of warfare, just like he had witnessed for himself during the First World War. Germany was leading the field in aircraft design and Herman Goering, Commander of the Luftwaffe, was overseeing a massive programme of aircraft development and production. Major cities like London, Coventry, Birmingham and Liverpool would be within reach of the German Air Force. With her strong navy and geographical location, invasion of the United Kingdom was unlikely without air superiority, and Goering realised that the German Luftwaffe would have to achieve air superiority before an invasion could be attempted. Matthews envisaged a vast airborne Armada comprising hundreds of bombers, flying at high altitudes, carrying thousands of bombs to be dropped on cities far below. Conventional air defences such as anti-aircraft batteries and barrage balloons, a fixture of city skylines throughout the war, were of limited value against the onslaught of the Luftwaffe. Ungainly and clumsy, barrage balloons provided only partial protection against enemy aircraft. Raised and lowered as required, they were only a threat to low- flying aircraft, their size and weight making them unsuitable for high altitudes and so offering no deterrent to bombers. They were also an easy target for Stuka dive bombers, whereupon a successful hit would turn the barrage balloon into a fireball that would fall on the very building it was meant to protect. It was extremely difficult for searchlights and anti-aircraft batteries to track fast-moving, low-level flying aircraft and barrage balloons were designed to force aircraft to higher altitudes so that they could then be targeted by anti-aircraft fire. But of course at higher altitudes they were not only out of the reach of searchlights but also a smaller target and therefore more difficult to hit. Barrage balloons were also vulnerable to bad weather and were known on occasions to break loose, posing a real threat to power cables. A skilful German pilot could weave his way through the mooring cables and just in case the pilot did have a mishap, the leading edge of aircraft wings had a device that could cut through the balloon's mooring cable. Some barrage balloons did have innovations which included mooring cables with explosive sections that were triggered when snagged by the wing of an aircraft, making them a more credible threat. The use of anti- aircraft guns to shoot down high-flying bombers was also a rather limited answer to such a threat, because simply firing millions of shells into the air in the hope of hitting a target, forced to fly at higher altitudes by barrage balloons, brought its own danger for those shells would only rain down back to earth.

Matthews was all too aware of Britain's precarious anti-aircraft defences and was convinced that gunfire and barrage balloons just weren't the answer to the threat of a modern air force whose fighters could fly in excess of speeds of 400mph and whose bombers were able to reach altitudes of 30,000 feet, higher than Mount Everest. His solution to the problem was an aerial defence scheme that involved creating an aerial minefield, up to a height of 30,000 feet, that could be quickly deployed above any city or indeed above anything that the Germans saw as a target.

A small rocket launched from the ground would, at a predetermined height, fire a series of smaller secondary rockets in all directions 'like the spokes of a wheel'[33] to a distance of about 200 yards. These secondary rockets would be released, by a timing fuse, either all at once or at different heights as the parent rocket gained altitude. Inside each of the smaller rockets was a carton that contained a small bomb attached to a parachute by a length of serrated wire. As the carton fell back to earth the tiny parachute would be deployed with the small bomb attached to it with the whole thing falling slowly to earth thereby creating an aerial minefield. Able to reach far greater heights, more quickly, and cause more damage than artillery shells, this scheme offered a real solution to the danger of aircraft attack. The serrated edges of the wire would act like a saw when snagged against the wing or propeller of an attacking aircraft, with the bombs exploding on impact with the aircraft. The use of a reverse time fuse would prevent the bomb from exploding if it reached the ground without hitting an enemy aircraft. Meanwhile the parent rocket, having reached the end of its climb, would deploy its own parachute, stowed away in the nose cone, and so return to earth where it could be easily retrieved and later reused. Matthews said that 'if bomber squadrons come over in cloud layers, they can be picked up by sound locators on the ground, and their range and speed accurately determined. The rocket torpedoes will then be shot from the ground, their fuses timed for the correct heights, aimed at the bracketing above and below the bombers, thus creating a minefield around them at a moments notice.'[34] Rockets have a distinct advantage over bullets or shells whose velocity soon begins to slow, whereas a rocket's velocity only increases the higher it climbs. Also his rocket, with a reach of 200 yards at any given altitude, covered a much wider area than an exploding shell ever could. He suggested that mobile units could be set up, with his rockets ready armed, so they could be deployed quickly and efficiently. The rocket wasn't just confined to land, for he saw no reason why it couldn't be easily launched from the deck of a ship. Matthews had spent some time in Germany were he carried out 'investigations in connection with rockets'.[35]

This was truly a remarkable concept and Matthews, with all his persuasive powers, managed to interest two wealthy industrialists, Sir Hugo Cunliffe-Owen and Sir James Hamet Dunn, in his aerial defence scheme. Sir Hugo was the chairman of the British-American Tobacco Company and had worked for the Ministry of Information during the First World War. Sir James, a close friend of Lord Beaverbrook, was a Canadian financier, stock broker and merchant banker. Both were initially prepared to finance Matthews who was then collaborating with a well-known German rocket engineer, Friedrich Wilhelm Sander. Sander had worked with the celebrated German scientist Fritz Von Opel to make both the world's first rocket-powered car and a rocket plane. During the 1930s Von Opel had begun to covertly manufacture rockets that had military applications. He had links with foreign governments who were interested in his work and the Nazi party began to get suspicious of him. He was arrested and died in prison in 1938. One of Sander's overseas contacts was Matthews, who invited him to Tor Clawdd to discuss the arrangements for building the aerial rocket which, claimed Matthews, cost 'one tenth of the ordinary shell'.[36] When Matthews visited Sander in Germany he witnessed the production of rockets based on the principle of his aerial torpedo and the preparations the Germans were making to 'prevent the penetration of their defences by hostile aircraft'.[37] They collaborated over the design and arrangements were made for it to be built. Sander visited Tor Clawdd to make arrangements for the delivery of the necessary equipment, and foundations to house the equipment were laid at Tor Clawdd. Back in Germany rockets were being manufactured and tested at Peenemünde. With extensive testing facilities, workshops and laboratories Peenemünde was home to thousands of rocket scientists including the talented Wernher von Braun. After the war von Braun would work for NASA and was instrumental in helping Neil Armstrong and the Apollo 11 crew to walk on the moon in July 1969. Both before and during the Second World War Peenemünde was the hub of the German rocket programme and was where the dreaded V2s were manufactured.

But before the arrangements were completed Sander was arrested by the Gestapo who had learned of the rocket scientist's dealings with Matthews, a British subject. According to Barwell the British Secret Service was also interested in the activities of Sander and a secret agent who was investigating his activities in Germany went missing. Hearing of all this, Sir Hugo and Sir James became nervous and after meeting with Matthews decided to pull out, leaving him to carry on under his own resources. But despite the fact that he was on a government blacklist they did express an interest in his aerial defence scheme. Major Wimperis, who knew of Matthews through his dealings with his electric beam during the 1920s, asked if 'a demonstration of such rockets can be offered ... so that the Air Ministry could consider its adoption for use in Air Defence'.[38] Shortly after Matthews visited the Master General of the Ordnance, in January 1936, to explain his aerial torpedo and he received a letter informing him that if he was 'able to demonstrate what you claim we shall definitely be interested'. However they were 'unable to bear any expenses in connection with the production of the quantities of the propellants required for the initial tests, but [could] provide the necessary facilities ... to enable the trials to be made'.[39] The government was certainly interested in his idea but Colonel L.V.S. Blacker, an enthusiastic advocate of Matthews' aerial defence system warned him of the opposition he would face in trying to get his system adopted. What was particularly brilliant with the scheme was that it would have provided an effective defence against the then unknown V1, or Doodlebug, and V2 rockets that would pose a dramatic and savage threat to Britain during the latter stages of the war. Sleek and menacing, the V1 and V2, Hitler's vengeance weapons, were launched from specialised launching sites and once airborne flew a straight, and therefore predictable, course. With this knowledge the aerial rocket could be fired to within very close proximity of the missiles with a good chance of its cluster rockets hitting them. With the V1 able to travel at speeds of 600mph and its successor faster than the speed of sound, the Royal Air Force's fastest fighter, the Spitfire, with a top speed of 460mph, was just too slow and struggled to intercept them.

The V1 or 'Doodlebug' was first used in June 1944 and 8,000 were launched over the next 80 days with 2,300 reaching London, causing a terrifying level of destruction.[40] The government's response was 'Operation Crossbow' which involved the deployment of numerous anti-aircraft batteries at strategic sites along the south coast of England. Hundreds of barrage balloons were also deployed in the rather vain hope that the V1s would become snagged against the mooring cables, but like German fighter planes the wings of the V1 had cable cutters enabling them to slice through any cables that obstructed their flight path.

Matthews did manage to give a successful demonstration of his aerial torpedo at the Calvary Club where 'even the more sceptical were convinced that there was something vastly superior in air defence to anything conceived'.[41] Colonel Norman Thwaites, Secretary of the Air League, noted that 'It will be deplorable if Grindell Matthews' device is not immediately adopted for energetic trial.'[42] Colonel Etherton wrote to Matthews, 'I feel that you are on the right lines and that you have gone further than any other inventor. If you get the full support for your scheme which it merits, then we shall be able to regard the possibility of foreign attack from the air with comparative equanimity.'[43] But even though Matthews had given a demonstration it was, as with all the previous inventions he took to the government, shelved – and why that was the case remains a mystery, for it was simply brilliant. In the summer of 1941, shortly before he died, Matthews made a visit to 10 Downing Street to discuss the details of his 'Aerial Torpedo' with Major Desmond Morton.[44] After submitting detailed drawings and charts of his scheme he returned to Swansea only to receive a letter a short while later thanking him for his plans but stating that they had no immediate plans to adopt his scheme. Mr D. Owen Evans MP, of the Mond Nickel Works in Clydach, offered to help finance Matthews' experiments with his aerial minefields. But even he struggled to get the necessary materials and supplies and failed to interest other parties.

Reports started appearing in a National Newspaper stating that

> Berlin reports that Britain is using a new anti aircraft device. A long range Focke-Wulfe bomber attacked a British ship in the Atlantic. The vessel fired eight rockets which released cables attached to a parachute and an explosive device. One of these wrapped around the wing tip of the aircraft, detonating when dragged free by the slipstream![45]

Had the government 'stolen' Matthews' idea? Dr A.D. Crow, working for the government, and knighted for his efforts, oversaw the development of 'Z' guns as an anti-aircraft weapon. This was a solid fuel rocket system – Matthews' system used liquid fuel – but it was clearly related to his original aerial minefield scheme. It wasn't just Britain who saw the future of rockets in warfare: by January 1940 German forces were using rockets to deliver explosives.[46]

Early in 1934 Matthews made the acquaintance of Mr P.E. Cleator, who was at the time 'struggling to establish the then infant [British] Interplanetary Society'.[47] They became firm friends, with Cleator holding Matthews in 'high regard' and visiting Tor Clawdd on several occasions where they discussed rocket research and design. Cleator recalled that he saw what he thought to be 'concrete observation trenches'[48] but these may well have been the foundations Matthews had installed, when collaborating with Sander, to house his rocket system to launch his aerial minefield. He told Cleator that another world war was close at hand and rockets would play an important part. Matthews asked Cleator if he would consider moving into Tor Clawdd to work with him. 'It was a tempting offer, which I declined on the grounds that I was not interested in the rocket as a weapon of war.'[49] Matthews became a member of the British Interplanetary Society (BIS) in July 1934, the same year as Professor A.M. Low. Low was an amazing inventor with dozens of patents to his name. A biography, *He Lit the Lamp*, provides details of this remarkable man.[50] The quintessential, absent-minded English professor, Low was born in 1888 in Purley, London. After finishing college Low worked for his uncle's engineering firm and had his first commercial success with a gadget called 'The Chanticleer' which was a whistling egg boiler and sold very well. Then, just before the First World War broke out, he invented a device he called 'Televista' which transmitted images by wireless. During the First World War he researched the possibility of guided missiles and embarked on a top secret project code named 'Aerial Target' which had a number of features, the like of which had never been seen before, including a compressed air launching mechanism and a guidance system that used an electrical gyroscope. But, in spite of this, just like the experience Matthews had with officials, the project was thwarted by a complete lack of vision on the part of government and the whole scheme was dropped. After the war he set up the 'Low Engineering Company Ltd' where he continued his passion for inventing. But like Matthews he was useless when it came to business matters, being only interested in technical matters of invention, and financial success eluded him.

When Matthews and Low joined the BIS they 'gave invaluable encouragement and assistance in many ways, at a time when association with us earned nothing but derision and abuse'.[51] The BIS, with its headquarters in London, is 'the world's longest established organization devoted to the promotion of astronautics'[52] and it was through his association with the BIS that Matthews became interested in the technology of space travel and convinced that man would, in the future, travel into space. Together with Cleator he explored the possibilities of using liquid hydrogen to power a rocket-plane to a speed of six miles per second. But Matthews' work in rocket research was dismissed by other scientists including Professor Lindermann, who was also a member of the BIS, who gave him 'short shrift' regarding his work in rockets. Lindermann also had a 'dismissive attitude to wartime reports of German V-2 activity'.[53] To the very end Matthews was ridiculed and dismissed, but an article he wrote, 'Travel by Rocket in 10 Years', appeared in the *Daily Mail* on October 29th 1936 and was remarkable in its predictions about the future of travel:

> By the proper use of rockets we shall be able to breakfast in London, lunch in New York, and be back in London in time for Dinner ... already projectiles are being fired by rocket propulsion to a distance of four and five miles. I have myself seen 18-pounder projectiles fired accurately at that range ... Travel in the stratosphere by rocket-plane will be anything from one to two thousand miles per hour ... passengers sitting in their hermetically sealed cabin ... the machine would be little different in appearance from the aeroplanes of to-day, except that there would be no propeller ... there will be no engine, as we know it, but in each wing will be three or four repulsors, all of them capable of developing 200 h.p. A special liquid fuel will take the place of petrol ...'

This was the 'Stratoplane' and the future of air travel as Matthews saw it in 1936.

Although exiled from the scientific establishment, ignored by the British government and with his finances at a low ebb, Matthews still kept on inventing during the late 1930s. Barwell reports[54] that he invented a device that could remotely explode magnetic mines. Consisting of 'a number of turns of wire in the form of a loop, which can be fitted to an aeroplane or any other type of aircraft. The machine would fly over the shipping lanes, and the loop would have suitable current fed into it from a small generating set. The effect would be the mines would detonate.' He never lost his interest in cinema and looked into the possibilities of making three-dimensional pictures. Working with various arrangements of projecting lenses, 'He had some success, but lack of finance prevented him from developing it as he would have liked.'[55] He also had an interest in biochemistry where he studied the action and nature of enzymes, biological catalyst that break down large complex molecules into small, simpler molecules. He found that he could, after some trial and error, crystallise enzymes.

On September 3rd 1939 whilst at Tor Clawdd, Matthews heard Neville Chamberlain announce that Britain was at war with Germany. He had been expecting it for months and wasn't convinced that a looming conflict would be averted when the Prime Minister was seen waving a piece of paper and proclaimed 'peace for our time' after meeting with Hitler to negotiate a European peace deal. Between 1934 and 1941 Matthews had made regular trips to Germany to meet with colleagues and other scientists working in the same fields as he was. He was aware of how Germany was gearing up for another major conflict. During his close discussions with his German friends he was only too aware of plans to make terrifying new weapons.

But by now Matthews was running into serious financial difficulties, 'was forced to sell some of his priceless machinery'[56] to settle his bills, and was very grateful and relieved to receive an offer from the Pacent Engineering Corporation in New York. Founded in 1919 as the Pacent Engineering Company, the company was incorporated in 1933 as a consulting engineering firm.[57] Louis Pacent was well acquainted with the achievements of Matthews and thought 'his life as a film story might have considerable interest to certain American film companies'.[58] Details of the offer the Corporation made to Matthews are not known but Louis wrote to him say 'we can use you over here, G.M., and I shall be glad if you will come over as my guest, all facilities will be provided'.[59] But it wasn't to be.

Notes

1. Conversation (30/12/07) and e-mail (02/01/08) with Mrs Helen MacDuff.
2. *Practical Mechanics*, October and November 1935.
3. Ellis, Ken and Jones, Geoff, *Henri Mignet and his Flying Fleas* (Haynes Publishing Group, 1990).
4. Practical Mechanics, October 1935.
5. Undated newspaper article 'Y Llais', in possession of Jeff Morgan.
6. www.tytwp.plus.com/Waun/Clydach2.html#top (accessed 27/12/07).
7. Conversation with Gari Melville, 16/03/07.
8. National Archives ADM 116/4766.
9. Barwell, p. 64.
10. Letter to the Admiralty dated February 4th 1931. National Archives ADM 116/4766.
11. Ibid.
12. Ibid.
13. National Archives ADM 116/4766.
14. Ibid.
15. Barwell, p. 163.
16. National Archives ADM 116/4766.
17. Letter from Messrs Sheard, Breach and Co. to Admiralty. National Archives ADM 116/4766.
18. National Archives ADM 116/4766.
19. Letter from Messrs Sheard, Breach and Co. to Admiralty. National Archives ADM 116/4766.
20. National Archives ADM 116/4766.
21. Ibid.
22. Ibid.
23. *Oxford Dictionary of National Biography* Vol. 28.
24. Barwell, p. 143.
25. Ibid., p. 144.
26. Ibid.
27. National Archives ADM 116/4766.
28. Ibid.

29. Ibid.
30. Ibid.
31. Letter dated September 8th 1936. Swansea County Archive D/DZ 346/13.
32. Ibid.
33. Barwell, p. 137.
34. Ibid.
35. Letter dated April 15th 1937 from Matthews to Colonel Buist. Swansea County Archive D/DZ 346/14.
36. Ibid.
37. Undated document marked 'confidential'. Swansea County Archive D/DZ 346/13.
38. Letter dated June 11th 1935 from H.E. Wimperis to Col. P.T. Etherton. Swansea County Archive.
39. Letter dated February 14th 1936, from Major-General Lewis to Matthews. Swansea County Archive D/DZ 346/13.
40. Smith, Peter J.C., *Air-Launched Doodlebugs. The Forgotten Campaign* (Pen and Sword Books Limited, 2006).
41. Letter from Colonel N.G. Thwaites to Lestock, dated May 28th 1937. Swansea County Archive.
42. Ibid.
43. Letter from Colonel Etherton to Matthews, dated 1937. Swansea County Archive.
44. Undated document by Arthur Henderson held in the Swansea County Archive, D/DZ 346/14.
45. Ibid.
46. 'Is Super Rocket Secret Weapon?' by Harry Gregson. Article in *Everybody's Weekly*, quoted in Barwell, p. 164.
47. Letter from P.E. Cleator to Mr Lloyd dated March 19th 1987.
48. Cleator, P.E., *Into Space* (Allen & Unwin, 1954).
49. Ibid.
50. Bloom, Ursula, *He Lit the Lamp. A Biography of Professor A.M. Low* (Burke Publishing Company Ltd, 1958).
51. Cleator, *Into Space*.
52. www.bis-spaceflight.com/HomePage.htm (accessed 28/12/07).
53. 'Terminal Testimony', *Journal of British Interplanetary Society*, 1986.
54. Barwell, p. 164.
55. Ibid.
56. Ibid., p. 166.
57. www.pacentengineering.com/history.htmhttp://bt.my.yahoo.com/ (accessed 28/12/07).
58. Letter dated September 30th 1941 from Louis Pacent to Ernest Barwell. Swansea County Archive D/DZ 346/10.
59. Barwell, p. 69; 'Ray genius who loved his fellow men' by Ron Vince, *Western Mail* 1972, Swansea County Archive D/DZ 346/11.

CHAPTER 10

The Final Frontier

On the morning of September 11th 1941, Lilley went into the lounge with Matthews' tea tray only to discover him slumped, motionless, over his writing desk. Unable to rouse him and panic-stricken, she phoned for the doctor. Arriving a short while later Dr Williams examined Matthews' motionless body and pronounced him dead. His death, announced in *The Times* obituaries on the following day, was the result of a coronary thrombosis and atheroma or 'furring' of the arteries. Six days after he died, on September 17th, Matthews' body was cremated at Pontypridd, 12 miles north of Cardiff. After the service, which was conducted by Rev. Christopher, Alfred, his brother, took away his ashes and scattered them over the mountainside at Tor Clawdd. There were only five people at his funeral: William Richards, Doug Smith (William's son-in-law), Matthews' brother, Alfred, P.C. Jack John and the undertaker. Over the years Matthews had lost touch with his brother and Edna Bodycombe remembers how cross Alfred felt that he hadn't got in touch with his younger brother. The arrangements to the funeral were very low-key, secretive even.

Shortly after the inventor's death, his biographer, Ernest Barwell, went up to Tor Clawdd to retrieve his diaries and some of his personal papers but noticed that many of his notes and papers were missing and much of those that remained had been censored. Pages had been ripped out, unlikely to have been done by Matthews who painstakingly kept meticulous records throughout his life.[1] Had the government removed his papers? Matthews had destroyed a lot of his own files and apparatus shortly before his death as he was, at that time, planning to move to America to work for the Pacent Engineering Corporation and wanted to avoid the possibility of his work falling into enemy hands. What equipment remained was collected by his old assistant, Lynes, who by that time had his own electrical firm in Middlesex. So whether the government had a hand in the disappearance of some of Matthews' work is a matter of speculation. The fact that things were missing hints that someone thought something was important. Considering what Matthews was doing and the reputation he had, particularly in Whitehall, it is not unreasonable to think that they had a hand in it. It was rumoured at the time[2] that government officials had entered Matthews' laboratory and taken away piles of papers, documents and various items of equipment. There was no correspondence about Matthews dealing with Goebbels and the Germans, so perhaps that was taken away. This has echoes of Nikola Tesla's treatment at the hands of the American government in 1943 when he died. Documents show that in 1940 Matthews was making arrangements to lease Tor Clawdd to an unnamed party in preparation for his move to America.[3] But in the event of his death Tor Clawdd and the remaining contents, including the furniture, was sold off by Messers Hobbs and Jones of Swansea for the sum of £500.[4] However his contact details were to remain in the local telephone directory until 1944. First entered in 1935, just after he moved in, was 'Grindell-Matthews H, Engineer, Tor Cloud [sic], Caig Cefn Parc Swansea, telephone Clydach 18'. However his number was to change in 1939 to Clydach 2118. But then there is a twist. He is still registered in the directory, under his name and occupation, until 1944, three years after he died. There must have been someone living there if the phone was still connected and the bill was being paid. Or it could simply be that the telephone directory wasn't updated as quickly as it should have been – after all there was a war on.

So what are we to make of Harry Grindell Matthews? Although there is no memorial, no awards or streets named after him and no blue plaque, every time a mobile phone is used, a movie is watched, a radio is switched on, or a blip appears on a sonar screen, Harry Grindell Matthews lives on through his achievements. One thing is certain: history has not been kind to him and to this day he is still dismissed by some as a crank and a fraudster. But so much evidence points to the contrary. Those who worked with Matthews found life exciting and full of possibilities. Lynes, his loyal assistant, had gone to London to run a successful engineering firm and many of those who had worked closely with him had gone on to build successful careers in the electronics and aviation industries. Both Jim Hill and Gwillym Price whom he helped build the 'Flying Flea' back in the mid-thirties, got employment with the Westland Helicopter Company in Yeovil. George Howell, who helped build the 'electric beam', went on to work on radar for the RAF. Official secrecy has meant much of the work he undertook has remained unknown and unacknowledged. He left no fortune: in fact he had to sell some of his equipment to pay his more pressing bills.

This account of Matthews' life is not an apology for a forgotten inventor with a reputation as fraudster and a crank, but a serious attempt to break through the myth surrounding him. Documents held in archives show that Matthews was indeed a pioneering inventor, living in an age of cynicism, presided over by short-sighted governments. With contradictory dates, omissions and misrepresentations it's not difficult to see how the myth of 'Death Ray Matthews' was born and why it still persists to the present day. But Matthews wasn't the only inventor to be on the receiving end of official bureaucracy and short-sightedness: Frank Whittle, R.J. Mitchell, Barnes Wallis and John Logie Baird all had to endure a lack of vision from the governments of the time. But inventors don't invent for riches or fame but because they think they can make a real difference and a worthwhile contribution, particularly during times of conflict.

There has been some considerable over-dramatisation of the Grindell Matthews story with allegations that he never actually made a 'death ray'. He did however invent an electric beam. With the dramatic and exaggerated claims that appeared in the press, his electric beam was never going to live up to those claims. He definitely had it, the government saw it and there were eyewitness accounts of him using it on the Mynydd y Gwair Mountain. The demonstration was never going to impress those government officials who went to his Harewood laboratory to witness it in action. The 'death ray' film he made was to promote his electric beam, hinting at the possibilities if only he could develop it. Ultimately his work with the electric beam took Sir Robert Watson-Watt down a road that led to radar.

As for wireless communication, he knew the future direction of this burgeoning technology and was involved from the earliest days. His Aerophone was a truly remarkable invention that ushered in a new era of communication when he transmitted the world's first press message in February 1912 whilst Marconi was still only transmitting Morse code. It's all too easy for his critics to overlook his achievements in radio telephony, submarine detection and talking films and point to the 'death ray'. Why does Marconi overshadow the history of wireless communication and why is he considered 'The Father of Wireless' when Fessenden beat him to the world's first radio broadcast only to be closely followed by Matthews?

If endeavour and not money or fame was a measure of success then the life of Harry Grindell Matthews is one of resounding success. Riches would have surely been his if the film industry had had the foresight to have seen where the future of the industry lay. Although he had an eye for a publicity stunt, he was by temperament quiet and unassuming, a gentleman. He didn't look like the archetypal inventor but was always immaculately dressed with meticulous habits. What is strange about the story of Matthews is the lack of acknowledgement in books about the history of wireless or films – they don't give him a single mention. No biography of Marconi gives him a mention and surely this prominent inventor knew of his work, but in his anxiety to claim the crown of 'Father of Radio' Marconi never spoke of him. Marconi's achievements in the field of radio are remarkable but Matthews' work was equally, if not more, important. Matthews does get a mention in *Who was Who, 1941–1950* as 'occupied in research [and] engaged upon methods of Defence from Air Attack' and is mentioned in biographies of Baird and Tesla. There is an excellent article about Matthews available on the internet[5] and a drama documentary[6] was made about him in the early 1990s but all too often he is a footnote, an afterthought.

Did Matthews undertake work for the government with defence weapons? It is known that he did work for them during the First World War in wireless control systems. Although it is widely accepted that he was rejected by the government, a military presence was placed at Tor Clawdd and he apparently had a bodyguard for a while. The government's work in the wireless transmission of power and rockets bore an uncanny resemblance to his own work. In 1942 American soldiers from the 2nd Infantry Division were stationed at Tor Clawdd, and set up camps with firing ranges as part of later preparations for the D-Day landings. The reason for them being stationed there is unclear. There were no facilities, it was bleak and miserable: why not be stationed somewhere else such as Ammanford? Were they placed there to protect the lab? There are several bomb craters in the locality. Were these the result of a determined effort by Germany to destroy Tor Clawdd? After all the German authorities were aware of its existence because German officials had made several pre-war visits when they were interested in the 'death ray', Sky Projector and Matthews' research into rockets. Even Dr Goebbels, the Nazi propaganda minister, was interested in the Sky Projector, wanting it for Nazi rallies. After Matthews' death it is now widely thought that American officers used Tor Clawdd as an officer's mess and this would explain why his telephone line was still connected throughout the war and after his death. Matthews had, in 1937, travelled out to Germany to meet with officials to discuss the possible sale of the Sky Projector and the present position on rocket research. Apparently in his final days he was approached by a gentleman from the German embassy in Dublin saying that the Nazi party was prepared to offer him £50,000 for the 'death ray' secret but Matthews, suspecting he was negotiating on behalf of the Nazi party, refused to sell. A letter[7] from the Army Department in Washington stated that they have no record of American soldiers being stationed at Tor Clawdd between 1941 and 1942. But they were there – the whole village knew about them – and today, over 60 years later, amateur metal detectors unearth the spent bullets fired by those American soldiers stationed on the Mynydd y Gwair Mountain.

It had been suggested that the £25,000 awarded to Matthews for *Dawn* was awarded improperly. This simply wasn't the case. Although the Admiralty had to approve the final payment, Lord

Fisher, a formidable personality, had a reputation for not suffering fools gladly had the utmost respect for Matthews and his work. £25,000 was an enormous sum for 1915 and showed how seriously the British Government took his work. He worked on the Aerophone for nine years, and submarine detection for 17 years: he was no conman. His work on the recording of sound on film between 1919 and 1923 was truly remarkable and demonstrates his dedication to the art of invention. Those that met and knew him describe him as modest and intelligent, 'a real gentleman'. Matthews' life was hectic and quite bewildering in its diversity.

Harry Grindell Matthews was not a man trying to make easy money by duping gullible investors. He never made money for himself, quite the contrary – he died in relative poverty, exiled from the British scientific establishment. He only wanted financing so that he was free to carry on inventing. For most of the first half of the twentieth century he was headline news around the world, a master at generating publicity to further his inventions, but was totally inept in all business matters, all too apparent with the failure of his many business ventures. Did he set them up to milk them of money? Well the bankruptcy papers detailing his creditors and the items he bought read like a shopping list for a laboratory, all of which was bought with the money of investors, who received a stake in his patents. The dusty, yellowing papers held in the National Archives are witness to a man who was an inventor and a true pioneer in an age of scepticism. Not an easy man to live with, he did have his dark side. Irascible, introverted, secretive and obsessive, his work totally ruled his existence which goes some way to explaining why his three marriages failed.

Most people seemed to like him. He instilled loyalty in his employees, many of whom went on to get important jobs in the electronics industry. He did little to dispel the myths surrounding him, it enabled him to work with a continuing degree of elusiveness so that he, and only he, knew the true extent of the work he undertook. He could be very charming and kind and those who knew him remember his intellect and warmth.

The story of Harry Grindell Matthews is one of endeavour, hope, inspiration and ultimately triumph, for he never gave up.

It's difficult to imagine life without the myriad of wireless gadgets we enjoy today: television, mobile phones, the internet, satellite navigation, fax machines and so on. Take them away and life comes to a shuddering halt. This is the man who was told by Winston Churchill that he was '100 years too advanced'.[8] Matthews laboured for others to reap the rewards, never to grow bitter – disappointed yes, but he was happy with the drama of it all and content to live the life he'd chosen. There is still some mystery surrounding Matthews and facts about his work and relationship with the British Government may still be uncovered, but this book has attempted to fill in some of the major gaps. For all the mystery and intrigue one thing is certain: for all his achievements, recognised or not, Harry Grindell Matthews was truly a man ahead of his time and deserves to be remembered.

Notes

1. Conversation with Gari Melville, 16/03/07.
2. Conversation with Edna Bodycombe, 15/01/07.
3. Papers held by Gari Melville.
4. Swansea County Archives D/D Z 346/11.
5. www.forteantimes.com/features/profiles/193/grindell_death_ray_matthews.html (accessed 20/08/07).
6. 'Y Dyn ar y Mynydd' [The Man on the Mountain], Dilyn Ddoe, broadcast by S4C, March 20th 1999.
7. In possession of Gari Melville.
8. Conversation with Edna Bodycombe and Jeff Morgan, 15/01/07.

APPENDIX I

Time Line

1880
March 17th — Born Winterbourne, Gloucestershire.

1888 — Starts at Mrs Webb's School, Alveston, South Gloucestershire.

Circa 1893 — Attends Merchant Venturers' College

1896 — Apprenticed to a Bristol electrical engineering firm.

1899 — Joins Baden-Powell South African Constabulary and serves in the Boer War.

1902 — Invalided home from South Africa and becomes consulting engineer for Lord De La Warr. Makes first radio broadcast from the Kursaal in Bexhill-on-Sea. Researches possibilities of wireless-controlled torpedo.

1904 — Marries Katy Williams.

1907 — Transmits speech by radio waves from the roof of the Kursaal, Bexhill-on-Sea where he establishes a radio station.

1909

November — Files 1st patent: '*Improved Means for Effecting Telephonic Communication Without Connecting Wires*'.

December — Files 2nd patent: '*Improved Automatic Righting Device for a Flying Machine*'.

1910

March — Raymond Phillips files patent: '*Improvements in or connected with the Controlling of Aerial Vessels by Wave Transmitted Electricity*'.

April — Grindell Matthews Wireless Telephone Company Ltd formed to develop and market the Aerophone.

May — London Hippodrome demo of wireless-controlled flying model airship, operated by Mr Phillips.

Constructs wireless-controlled airship at the New Passage Hotel, Pilning.

July — Becomes a member of the Royal Institute of Great Britain.

1911

May — Files 3rd patent: '*Improvements in telephone instruments*'.

August — Files 4th patent: '*Improvements in Wireless Telephony*'.

September — Attempts radio communication between Beachley Pier to Old Passage Ferry Inn, a distance of three miles.

Establishes a radio link between two French villages, a distance of 670 miles.

	Plans to sell the Aerophone to the public. World first at Ely Racecourse in Cardiff when Matthews makes radio contact with aeroplane in flight.
October	Plans to transmit speech a distance of 40 miles.

1912

January	'Voice of the North Sea Ghost'.
	Sets up transmitting aerial on roof of *Western Mail* and Newport and Cardiff linked together by wireless telegraphy and wireless telephony.
February	Transmits the first-ever wireless press message from Newport to the *Western Mail* buildings in Cardiff.
	Transmits a radio broadcaster to a *Daily Express* reporter a distance of 110 miles from the Black Rock (River Severn) to Dalston in Hackney, London.
March	Files 5th patent: *'Improvements of high frequency waves'*.
	Wireless transmission between two cars in motion.
	Gives a demonstration of his wireless telephone to Lloyd George, the Chancellor of the Exchequer and Colonel Seely, Secretary of State for War.
July	Demonstrates Aerophone to King George V at Buckingham Palace.
	Refused licence by British GPO to transmit between Britain and France.

'Marconi Scandal' becomes public knowledge.

With finance from the Grindell Matthews Wireless Telephone Company, sets up two radio broadcasting stations: one in Northampton and one at Letchworth in Hertfordshire.

1914
January — Plans radio transmission of speech from Newfoundland, Canada across the Atlantic.

April — Grindell Matthews Wireless Telephone Company Ltd goes bankrupt.

Closure of Northampton and Letchworth experimental radio stations.

First experimental trial of a boat controlled by a searchlight at Edgbaston, in Birmingham.

Collaborates with Dr Edmund Edward Fournier D'Albe.

Collaborates with Lord Fisher to construct an aerial torpedo to provide defence against Zeppelins.

Fatal accident results in the aerial torpedo project being shelved.

1915
October — Agreement signed between government and Matthews for the trial of *Dawn*.

Selenium tested under four different conditions.

December — Demonstrates *Dawn* on Penn Pond in Richmond Park, London. Awarded £25,000.

APPENDIX

	Works on wireless operation of mine exploder, bomb release mechanism and torpedo control.
1916	Changes his name by deed poll to Harry Grindell Grindell.
1917 January	Writes to the Ministry of Inventions saying that 'I have recently devised a means of locating enemy submarines in a simple, rapid and effective manner.'
March	Trials his submarine detection apparatus in Portsmouth Harbour.
March/May	Second trial at New Passage in Bristol, England.
June	Third trial at Barry, Wales.
	Matthews travels to Cherbourg, France, to continue his work with submarine detection.
September	Inquiry held into the events that took place during the trials at Barry.
1918	Government drops Matthews' work on submarine detection.
1919	First experiments with recording sound carried out at New Passage Hotel.
1921	Moves to 2 Harewood Place, Hanover Square, London.
September	Records an interview with Sir Ernest Shackleton using his new camera.
1922 August	Files patent for '*Improvements relating to Photographic Sound Recording*'.

1923	
June	Patents device *'Electric Gas-discharge Tube for Photographic Sound Recording'*.
1924	*The Death Ray*, 'The Most Startling and Breath-Taking Motion Picture Ever Made!' shown at several London theatres.
March	Matthews and Captain Edwards visit the Air Ministry and speak to Wing Commander Bowen and Major Lefroy regarding why he can't give a demonstration of his electric beam.
May	Questions asked in the House of Commons about the 'death ray'.
	Matthews demonstrates his electric beam to government at Harewood Place.
	Further questions asked in the House of Commons about the 'death ray'.
	Matthews in Lyon.
October	Patent No. 606 260 filed for *'Projection à distance Phénomenès invisibles de haute fréquency électrique'*.
	Matthews makes the first of several journeys across the Atlantic to America to promote his talkies, 'death ray' and Luminaphone. Meets Lady Edison.
	Death Ray film released in America.
1925	
February	Sails from New York to Europe aboard the *Acquitania*. 'Famed inventor of "Death Ray"'. Plans to return to America permanently and set up a lab in New York.

	Meets American divorcee Olive Waite. Returns to Europe.
September	Marries Olive Waite in Scotland, returns to New York. Sets up a workshop in New York.
1926 March	Files patent GB272606: *'Improvements relating to Optical Projection'*.
1927	Second 'movie' of 'death ray' released in Britain by Pathé.
1928	In Long Island, New York.
1929 May	Employed in America as consulting engineer for Warner Brothers.
1930 May	Contract with Warner Brothers ends.
July	Returns to England.
August	Sells the rights to his Sky Projector to a company called 'G.M. Sky Projector'.
September	Is a director and consulting engineer of 'G.M. Sky Projector'.
	Olive Waite divorces Matthews.
	File patent for *'Improvements Relating to Apparatus for Producing Musical Sounds.'*
1931 September	Receives bankruptcy order.
October	Matthews' creditors hold first meeting.

December	Submits his Statement of Affairs to bankruptcy court.
1932	
January	Luminastra set up.
	Matthews' public examination at High Court of Justice, Bankruptcy Buildings, Carey Street, WC2.
May	Patents Sky Projector in France
December	Takes out second patent for improved version of Sky Projector.
1933	Makes acquaintance of P.E. Cleator who established the British Interplanetary Society.
1934	
March	Declared bankrupt.
May	Constructs Tor Clawdd.
1935	Helps build a 'Flying Flea'.
	Begins work on Aerial Torpedo to defend cities.
1936	
October	'Travel by Rocket in 10 Years', article written by Matthews for the *Daily Mail*.
1937	
July	Luminastra Company dissolved.
1938	
January	Marries Ganna Walska.
1941	Visits 10 Downing Street, London to meet with officials to discuss Aerial Torpedo.
September 11th	Collapses and dies from a heart attack.

APPENDIX II

Known Addresses of Harry Grindell Matthews

Date	Address
1881	The Grove, Winterbourne, Gloucestershire
1909	Friezwood, Rudgeway, Thornbury, Gloucestershire
1915	125 Pall Mall, London, SW
1918	7a Bickenhall Mansions, Gloucester Place, London
1921	2 Harewood Place, Hanover Square, London
1925	89 New Oxford Street, London
1926	50 Pall Mall, London SW1
1928	Long Island, New York
1930	101 West 55th Street, New York
1931	The Felix Hotel, Jermyn Street, St. James, London SW1
1932	Midland Bank Chambers, Bromley, Kent
1933	26 Devonshire Street, London WC1
1934	Tor Clawdd, Craig Cefn Parc, Swansea, Wales
1936	20 Bolton Street, London W1
1937	Torrington Square, Holborn, London

APPENDIX III

Patents Filed by Harry Grindell Matthews

Number	Publication Date	Title	Number	Notes
1	03/11/**1910**	Improved Means for Effecting Telephonic Communication Without Connecting Wires	GB190925639	Aerophone device
2	01/12/**1910**	Improved Automatic Righting Device for a Flying Machine	GB190928083	Automatic Pilot
3	14/03/**1912**	Improvements in Telephone Instruments	GB191112730	
4	08/08/**1912**	Improvements in Wireless Telephony	GB191118284	Aerophone
5	15/05/**1913**	Improvements in Means for Producing Electro Magnetic Waves of High Group Frequency and in the Application thereof to Wireless Telephony	GB191206486	Generating 'carrier waves'
6	12/02/**1914**	Improvements in Arrangements for Producing Electro-magnetic Oscillations particularly for use in Radio Telephony	GB191314927	
7	12/03/**1914**	Improvements for use in Wireless Telephony	GB191312076	Aerophone
8	12/03/**1914**	Improvements in Switching Arrangements for use in wireless Telephony and Wireless Telegraphy	GB191312157	

APPENDIX

Number	Publication Date	Title	Number	Notes
9	04/06/**1914**	Improved Calling or Alarm Apparatus for use in Wireless Signalling	GB191313065	Burglar alarm
10	14/11/**1923**	Improvements relating to photographic sound recording	GB206908	Sound recording
11	09/03/**1926**	Electric gas-discharge tube for photographic sound recording	**US**1575701	Sound recording
12	08/07/**1926**	Improvements relating to apparatus for producing Musical sounds	GB254437	Tone wheel
13	03/05/**1927**	Perfectionnements à la projection optique	**FR**620965	Sky Projector Mark I
14	17/06/**1927**	Improvements relating to optical projection	GB272606	Sky Projector Mark I
15	27/09/**1927**	Optical Projection	**CA**274165	Sky Projector Mark I
16	22/10/**1931**	Improvements relating to apparatus for producing musical sounds	GB359125	Luminaphone
17	14/06/**1932**	Optical Projection	**US**1862577	Sky Projector Mark II
18	28/11/**1932**	Perfectionnements a la projection lumineuse	**FR**736708	Sky Projector Mark II
19	12/04/**1934**	Improvements in or relating to optical projection apparatus	GB408406	Sky Projector Mark II

Endnote

Just before publication, on a cold, wintry day in Chepstow the author had the privilege of meeting the great-niece of Harry Grindell Matthews. Inside the elegant eighteenth-century building that houses the Chepstow Museum the author was delighted as Mrs Stephens, née James, warmly recalled the family memories of her great-uncle. Although very young when Matthews died, Mrs Stephens remembered his brother, Alfred, very well. What became apparent to the author from his conversation with Mrs Stephens was the warmth and affection she held for Matthews. A human and personal side of the inventor emerged: his love of family, friends and close colleagues; his brilliant intellect; his fondness for animals, music and languages. Despite his frustration at the lack of co-operation he received from the British Government he remained whole-heartedly dedicated to his work. Meeting a family relative took the author much closer to the subject of this biography and the experience was a memorable honour.

On March 17th 2009, a blue plaque was unveiled to Harry Grindell Matthews. Partly funded by the Winterbourne Parish Council, the plaque was placed on 'The Grove', Matthews' childhood home in Winterbourne, South Gloucestershire. Unveiled to a crowd of local residents, the blue plaque provides a lasting reminder to the life of this remarkable inventor.

Further Reading

Biographies

Harry Grindell Matthews

The Death Ray Man: The Biography of Grindell Matthews, Inventor and Pioneer. Ernest Barwell. Published by Hutchinson (1943).

Guglielmo Marconi

Signor Marconi's Magic Box. How an Amateur Inventor Defied Scientists and Began the Radio Revolution. Gavin Weightman. Published by HarperCollins (2003).

William Friese-Green

Friese-Greene. Close-up of an Inventor. Ray Allister. Published by Marsland Publications London (1948).

Nikola Tesla

The Life and Times of Nikola Tesla. Biography of a Genius. Marc J. Seifer. Published by Citadel Press, Kensington Publishing Corp. (1998).
The Fantastic Inventions of Nikola Tesla. Nikola Tesla. Published by Adventures Unlimited Press (1993).

Lee de Forest

Lee de Forest and the Fatherhood of Radio. James A. Hijiya. Published by Associated University Presses (1992).

John Logie Baird

The Secret Life of John Logie Baird. Tom McArthur and Peter Waddell. Published by Century Hutchinson Ltd (1986).
John Logie Baird. A Life. Anthony Kamm and Malcolm Baird. Published by National Museums of Scotland Publishing (2002).
Television and Me. The Memoirs of John Logie Baird. John Logie Baird, edited and introduced by Malcolm Baird. Published by Mercat Press Ltd (2004).

Professor Low

He Lit the Lamp. A Biography of Professor A.M. Low. Ursula Bloom. Published by Burke Publishing Company Ltd (1958).

Marie Corelli

The Mysterious Miss Marie Corelli, Queen of Victorian Bestsellers. Teresa Ransom. Published by Sutton Publishing Limited (1999).

Madam Ganna Walska

Always Room at the Top. Ganna Walska. Published by Richard R. Smith, New York. (1943).

Technical Books

History of Wireless

History of Wireless. Tapan K. Sarkar *et al*. Published by John Wiley & Sons, Inc. (2006).
The Guinness History of Inventions. G.I. Brown. Published by Guinness Publishing Ltd (1996).

Popular Science

Backroom Boys. The Secret Return of the British Boffin. Francis Spufford. Published by Faber and Faber (2003).

A Computer Called LEO: Lyons Tea Shops and the World's First Office Computer. Georgina Ferry. Published by Fourth Estate (2003).

The Speed of Sound: Hollywood and the Talkie Revolution, 1926–30. Scott Eyman. Published by Johns Hopkins University Press (1999).

History of Film

The Speed of Sound: Hollywood and the Talkie Revolution, 1926–1930. Scott Eyman. Published by The John Hopkins University Press (1999).

The Talkies: American Cinema's Transition to Sound, 1926–1931. Donald Crafton. Published by University of California Press (1999).

INDEX

Entries in **bold** are illustrated.

Adastra Incorporated, 170
Admiralty, 4, 18, 40, 41, 43, 64, 65, 68–72, 87, 88, 89, 91–96, 98, 99, 125, 131, 133, 138, 143, 196–199, 201, 202, 212, 218
Aerial Torpedo, 65- 67, 206–208
Aerophone, 2, 3, 31- 35, 40–44, 46, 47, 48, 52, 53, 56, 57, 72, 217, 219
Augustin, Louis Aime, (French inventor) 104
Automatic Pilot, 31

Baird, John Logie, (Scottish inventor) 7, 107, 108, 119, 177, 187, 216, 217
Bankruptcy, 56, 167, 173, 174
Barwell, Ernest, 10, 17, 28, 29, 30, 32, 58, 59, 60, 64, 73, 74, 93, 100–103, 119, 120, 141, 143, 151, 152, 153, 167, 177, 178, 183, 191, 192, 193, 201, 206, 211–214
Beit, Sir Otto, 35
Bexhill-on-Sea, 22, 23, 25, 26, 29, 42, 108
Bloemfontein, 19
Board of Invention and Research, 64, 68, 70, 73, 88, 89
Bowen, Wing Commander, 130, 131, 133
Branley, Édouard Eugène Désiré, (French physicist) 17
Bristol, 11, 12, 13, 15, 16, 18, 19, 28, 33, 34, 93–98, 105, 112, 141, 197
Bristol Tramways and Carriage Company, 16
Brixton Prison, 200
Buckingham Palace, 3, 53
Bullet Circuit Closer, 101, 102

Cambridge, 68, 142
Cardiff, 3, 45, 51, 52, 93, 182, 214
Churchill, Sir Winston, 25, 53, 64, 184, 187, 188, 199, 220
Citroën, 170
Cleator, P.E., 209, 210, 213
Colorscope, 167
Corelli, Marie, (novelist) 47, 59
Count Valambrosia, 169, 170, 172, 173
Craig Cefn Parc, 179, 180
Cunliffe-Owen, Sir Hugo, 206

D'Albe, Edmund Eward Fournier, (English physicist) 4, 62, 65, 69, 70, 73, 74, 102, 107, 108, 129
Dawn, 4, 61, 63–70, 72, 73, 86, 107, 129, 142, 218
De Forest, Lee, (American physicist) 17, 25, 48, 117, 118, 119
De La Warr, Earl, 22, 24, 26, 31, 32
Death Ray, 1, 5, 7, 28, 115, 121–124, 127, 129, 131, 133, 135–138,

INDEX

142, 144–150, 182, 186, 187, 188, 195, 216, 217, 218
Dickson, William, (inventor) 104
Ditcham, William, 48, 49, 56

Edison, Thomas, (American inventor) 9, 19, 21, 25, 104, 145, 148
Edwards, Captain, 130, 131, 135
Etherton, Colonel P.T., 185, 199, 201, 208, 213

Fessenden, Reginald Aubrey, (Canadian inventor) 17, 25, 29, 86, 217
Fisher, Lord, 4, 63–69, 72, 73, 86, 87, 89, 93, 101, 142, 173, 219
Flat Holm, 141
Flying Flea, 194, 195, 216
France, 3, 5, 6, 18, 20, 40, 46, 53, 55, 58, 95, 98, 101, 104, 134, 135, 137, 139, 140, 169, 174, 192
Friese-Greene, William, (English inventor)106, 107, 119

George V, King, 53, 72
George, Lloyd, 53, 54, 58, 105, 213
Goebbels, Dr. Paul Joseph, 6, 177, 195, 215, 218
Goering, Hermann, 177, 203
Grindell Matthews Wireless Telephone Company, 34, 35, 38, 42, 43, 46, 48, 49, 55, 56, 72

Haggard, Sir Henry Rider, (author) 47
Harewood laboratory, 110, 126, 129, 216
Henke, Herbert, 51
House of Commons, 138
Houston, Lady, 185, 199, 200, 201
Howell, George, 187, 197, 216
Hucks, B.C. (aviator), 3, 45, 46
Hydrophone, 86

Instone, Samuel, 141
Ireland, Commander, 67, 203

Jennings, Edward, 17, 19, 20, 24

Kerr, Admiral, 140, 141
Kinetoscope, 104, 105
Kursaal, 22, 23, 24

Leitz, Dr. Ernest, 168, 169
Letchworth, (radio transmitting station) 55, 57, 58
Lodge, Sir Oliver, (English physicist) 17, 18, 108, 128
London Hippodrome, 35, 36, 37, 39, 114
Loomis, Dr. Mahlon, 18
Low, Sir Archibald, (English inventor and engineer) 128
Lumière Brothers, 105
Luminaphone, 6, 167
Luminastra, 175, 176
Lynes, B.J., (Matthews' assistant) 61, 63, 89, 90, 92–95, 101, 109, 126, 128, 131, 132, 147, 200, 202, 215, 216
Lyons, Sir Joseph, 32, 43, 58

Marconi, Guglielmo, (Italian inventor) 1, 2, 3, 17–22, 24, 25, 29, 39, 41, 42, 43, 48, 49, 52, 54, 55, 108, 124, 131, 141, 152, 177, 217
Matthews, Charlotte (sister), 10, 11
Matthews, Daniel (father), 9, 10, 11, 100
Matthews, Harry Grindell, 2–28, 31–58, 61–73, 86–96, 98 -105, 107–119, 121, 122, 124–151, 167–175, 177, 179–220
Matthews, Jane Grindle (mother), 9
Maxwell, James Clerk, (Scottish physicist) 15, 25
McKenna, Reginald, 68, 70

Mignet, Henri, 195, 212
Ministry of Inventions, 87
Moon Element, 62, 65, 69, 71, 73, 129
Morris, Lilley, 47, 182, 189
Mynydd y Gwair Mountain, 179, 180, 181, 183, 194, 216, 218

New Passage Hotel, 38, 41, 61, 93, 95, 96, 97, 109, 110, 117
New York, 5, 6, 116, 118, 145, 147, 148, 149, 151, 167, 169, 170, 171, 194, 210, 211
Northampton, (radio transmitting station) 55, 57

Oxford, 18, 110, 126, 199, 212

Pacent Engineering Corporation, 211, 213, 215
Parsons, Charles, (English inventor) 63, 168
Patents, 1, 4, 32, 40, 44, 46, 48, 56, 106, 114, 115, 117, 119, 135, 145, 148, 150, 174, 209, 219
Philips, Raymond, 36
Photophone, 73, 117, 118
Pilning, 34, 38, 46, 96, 102
Popoff, Alexander, (Russian inventor) 18
Portsmouth, 71, 89

Radar, 5, 15, 55, 108, 131, 150, 186, 187, 188, 193, 216, 217
Raytheon Co., 188
Robinson, Sir Clifton, 11, 34, 43
Royer, Eugene, 135, 150
Rutherford, Sir Ernest, (New Zealand physicist) 142

Sander, Friedrich Wilhelm, (German engineer) 206, 207, 209
Savery, Jacob, 13
selenium, 61, 62, 64, 66–70, 72, 109, 113, 167

Shackleton, Sir Ernest, (Anglo-Irish explorer) 4, 111, 112, 117
Sky Projector, 6, 168–177, 188, 195, 218
Sonar, 86, 215
Sperry Gyroscope Company, 169, 173
Stella Fregelius, 47
Stratford-upon-Avon, 47
Stratoplane, 6, 211
Submarine Detection, 85, 86
Swansea, 1, 2, 7, 28, 29, 30, 59, 60, 89, 94, 102, 103, 119, 122, 141, 151, 153, 178, 179, 184, 187, 188, 192, 193, 197, 200, 201, 208, 212, 213, 215, 220

Talkies, 2, 4, 102, 104, 107, 115–120, 121, 147, 168
Tesla, Nikola, (inventor) 7, 17, 145, 146, 149, 150, 153, 177, 215, 217
The Grove, 10, 11, 14
Thompson, Sir J.J., (English physicist) 142
Tizard, Sir Henry, 186, 193
Tolkien, J.R.R., (author) 19
Tone Wheel, 167
Tor Clawdd, 6, 150, 180, 181, 182, 184, 185, 189, 191, 192, 194, 195, 196, 206, 209, 211, 214, 215, 218

U-boats, 85, 86
Usborne, Commander, 65, 66, 67

V1, Doodlebug, 207, 208
V2 Rocket, 207
Vitaphone, 116, 118

Waite, Olive, (Matthews' second wife) 119, 149, 174, 175
Walska, Ganna, (Matthews' third wife) 6, 189, 190, 193

Warner Brothers, 5, 116–118, 148, 149, 171, 174
Watson-Watt, Sir Robert, (Scottish electrical engineer) 131, 186, 187, 217
Western Mail, 51, 52, 59, 60, 213
Whittle, Sir Frank, (English engineer) 7, 203, 216
Williams, Katy, (Matthews' first wife) 25, 26, 149, 171
Wimperis, Major, 131–134, 186, 193, 207, 213

Winterbourne, 9–12
Wolffe, Jabez, 40, 41
Wright Brothers, (American aeronautical engineers) 203

Yeats-Brown, Commander, 96, 98, 103

Zeppelins, 4, 65, 67, 70, 73